Everyday Mathematics®

The University of Chicago School Mathematics Project

Differentiation Handbook

Grade 5

Chicago, IL • Columbus, OH • New York, NY

The University of Chicago School Mathematics Project (UCSMP)

Max Bell, Director, UCSMP Elementary Materials Component; Director, *Everyday Mathematics* First Edition; James McBride, Director, *Everyday Mathematics* Second Edition; Andy Isaacs, Director, *Everyday Mathematics* Third Edition; Amy Dillard, Associate Director, *Everyday Mathematics* Third Edition; Rachel Malpass McCall, Associate Director, *Everyday Mathematics* Common Core State Standards Edition

Authors

Amy Dillard
Kathleen Pitvorec

Common Core State Standards Edition

Rachel Malpass McCall, Rebecca W. Maxcy, Kathryn M. Rich

Technical Art

Diana Barrie

ELL Consultant

Kathryn B. Chval

Differentiation Assistant

Serena Hohmann

Photo Credits

Cover (l)Steven Hunt/Stone/Getty Images, (c)Martin Mistretta/Stone/Getty Images, (r)Digital Stock/CORBIS, (bkgd)PIER/Stone/Getty Images; **Back Cover** Martin Mistretta/Stone/Getty Images; **9** (l)Barrie Rokeach/Alamy, (c)Image DJ/Alamy, (r)Damons Point Light/Alamy; **23 33** The McGraw-Hill Companies; **Icons** (Objective)Brand X Pictures/PunchStock/Getty Images.

Permissions

Carl Sagan quotation 9806 from Dictionary of Quotations Third Edition, Newly Revised, reproduced by kind permission of Wordsworth Editions, LTD.

Deciding to Teach Them All, Tomlinson, C., Educational Leadership 61(2), © 2003, reprinted by permission. The Association for Supervision and Curriculum Development is a worldwide community of educators advocating sound policies and sharing best practices to achieve the success of each learner. To learn more, visit ASCD at www.ascd.org.

Gregory, G., Differentiated Instructional Strategies in Practice, p. 27, © 2003 by Corwin Press Inc., reprinted by permission of Corwin Press Inc.

This material is based upon work supported by the National Science Foundation under Grant No. ESI-9252984. Any opinions, findings, conclusions, or recommendations expressed in this material are those of the authors and do not necessarily reflect the views of the National Science Foundation.

everyday**math**.com

Send all inquiries to:
McGraw-Hill Education
STEM Learning Solutions Center
P.O. Box 812960
Chicago, IL 60681

ISBN 978-0-07-657648-7
MHID 0-07-657648-5

Printed in the United States of America.

3 4 5 6 7 8 9 RHR 17 16 15 14 13 12

McGraw-Hill is committed to providing instructional materials in Science, Technology, Engineering, and Mathematics (STEM) that give all students a solid foundation, one that prepares them for college and careers in the 21st century.

Contents

Differentiating Instruction with *Everyday Mathematics*®

Philosophy

> *Differentiation is a philosophy that enables teachers to plan strategically in order to reach the needs of the diverse learners in classrooms today.*
>
> (Gregory 2003, 27)

This handbook is intended as a guide to help you use *Everyday Mathematics* to provide differentiated mathematics instruction. A differentiated classroom is a rich learning environment that provides students with multiple avenues for acquiring content, making sense of ideas, developing skills, and demonstrating what they know.

In this sense, differentiated instruction is synonymous with good teaching. Many experienced teachers differentiate instruction intuitively, making continual adjustments to meet the varying needs of individual students. By adapting instruction, teachers provide all students opportunities to engage in lesson content and to learn.

Though students follow different routes to success and acquire concepts and skills at different times, the philosophy of *Everyday Mathematics* is that all students should be expected to achieve high standards in their mathematics education, reaching the Grade-Level Goals in *Everyday Mathematics* and the benchmarks established in district and state standards.

Everyday Mathematics is an ideal curriculum for differentiating instruction for a variety of reasons. The *Everyday Mathematics* program:

- begins with an appreciation of the mathematical sensibilities that students bring with them to the classroom and connects to students' prior interests and experiences;
- incorporates predictable routines that help engage students in mathematics and regular practice in a variety of contexts;
- provides many opportunities throughout the year for students to acquire, process, and express mathematical concepts in concrete, pictorial, and symbolic ways;
- extends student thinking about mathematical ideas through questioning that leads to deepened understandings of concepts;

- incorporates and validates a variety of learning strategies;
- emphasizes the process of problem solving as well as finding solutions;
- provides suggestions for enhancing or supporting students' learning in each lesson;
- encourages collaborative and cooperative groupings in addition to individual and whole-class work;
- facilitates the development and use of mathematical language and promotes academic discourse;
- provides teachers with information about the learning trajectories or paths to achieving Grade-Level Goals;
- highlights opportunities for teachers to assess students in multiple ways over time;
- suggests how students can demonstrate what they know in a variety of ways; and
- encourages students to reflect on their own strengths and weaknesses.

The purpose of this handbook is to provide ideas and strategies for differentiating instruction when using *Everyday Mathematics*. This handbook highlights differentiation that is embedded in the program and also points to features that can be readily adapted for individual students. The information and suggestions will help you use *Everyday Mathematics* to meet the needs of all learners—learners who need support in developing concepts, learners who need support in developing language proficiency, and learners who are ready to extend their mathematical knowledge and skills.

This handbook includes the following:

- a lesson overview to highlight the features that support differentiated instruction;
- general differentiation strategies and ideas for developing vocabulary, playing games, and using Math Boxes, as well as suggestions for how to implement the lessons to differentiate learning effectively;
- specific ideas for differentiating the content of each unit, including suggestions for supporting vocabulary and adjusting the level of games; and
- a variety of masters that can be used to address the needs of individual learners.

A Lesson Overview

Everyday Mathematics lessons are designed to accommodate a wide range of academic abilities and learning styles. This lesson overview highlights some of the strategies and opportunities for differentiating instruction that are incorporated into the lessons.

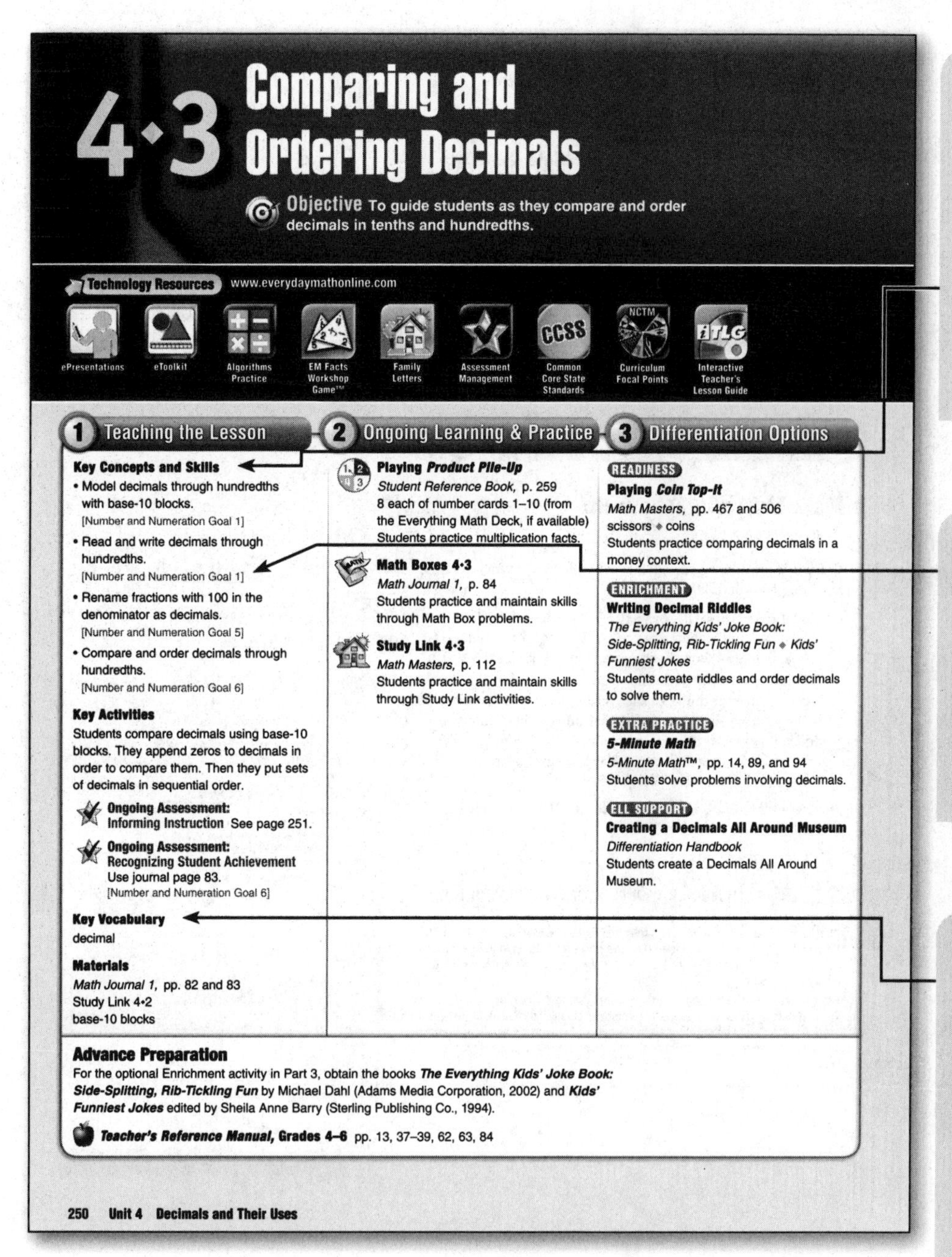

4•3 Comparing and Ordering Decimals

Objective To guide students as they compare and order decimals in tenths and hundredths.

Technology Resources www.everydaymathonline.com

ePresentations | eToolkit | Algorithms Practice | EM Facts Workshop Game™ | Family Letters | Assessment Management | Common Core State Standards | Curriculum Focal Points | Interactive Teacher's Lesson Guide

1 Teaching the Lesson

Key Concepts and Skills

- Model decimals through hundredths with base-10 blocks. [Number and Numeration Goal 1]
- Read and write decimals through hundredths. [Number and Numeration Goal 1]
- Rename fractions with 100 in the denominator as decimals. [Number and Numeration Goal 5]
- Compare and order decimals through hundredths. [Number and Numeration Goal 6]

Key Activities

Students compare decimals using base-10 blocks. They append zeros to decimals in order to compare them. Then they put sets of decimals in sequential order.

Ongoing Assessment: Informing Instruction See page 251.

Ongoing Assessment: Recognizing Student Achievement Use journal page 83. [Number and Numeration Goal 6]

Key Vocabulary

decimal

Materials

Math Journal 1, pp. 82 and 83
Study Link 4•2
base-10 blocks

2 Ongoing Learning & Practice

Playing *Product Pile-Up*
Student Reference Book, p. 259
8 each of number cards 1–10 (from the Everything Math Deck, if available)
Students practice multiplication facts.

Math Boxes 4•3
Math Journal 1, p. 84
Students practice and maintain skills through Math Box problems.

Study Link 4•3
Math Masters, p. 112
Students practice and maintain skills through Study Link activities.

3 Differentiation Options

READINESS
Playing *Coin Top-It*
Math Masters, pp. 467 and 506
scissors ◆ coins
Students practice comparing decimals in a money context.

ENRICHMENT
Writing Decimal Riddles
The Everything Kids' Joke Book: Side-Splitting, Rib-Tickling Fun ◆ *Kids' Funniest Jokes*
Students create riddles and order decimals to solve them.

EXTRA PRACTICE
5-Minute Math
5-Minute Math™, pp. 14, 89, and 94
Students solve problems involving decimals.

ELL SUPPORT
Creating a Decimals All Around Museum
Differentiation Handbook
Students create a Decimals All Around Museum.

Advance Preparation

For the optional Enrichment activity in Part 3, obtain the books ***The Everything Kids' Joke Book: Side-Splitting, Rib-Tickling Fun*** by Michael Dahl (Adams Media Corporation, 2002) and ***Kids' Funniest Jokes*** edited by Sheila Anne Barry (Sterling Publishing Co., 1994).

***Teacher's Reference Manual*, Grades 4–6** pp. 13, 37–39, 62, 63, 84

250 Unit 4 Decimals and Their Uses

Key Concepts and Skills are identified for each lesson and are linked to Grade-Level Goals. They highlight the variety of mathematics that students may access in the lesson and show that each lesson has significant mathematics content for every student.

Grade-Level Goals are mathematical goals organized by content strand and articulated across grade levels. These goals define a progression of concepts and skills from Kindergarten through Grade 6.

Key Vocabulary consists of words that are new or unfamiliar to students and is consistently highlighted. Students, including English language learners, are encouraged to use this vocabulary in meaningful ways throughout the lesson in order to develop a command of mathematical language.

Mental Math and Reflexes problems range in difficulty, beginning with easier exercises and progressing to more-difficult ones; levels are designated by the symbols ●○○, ●●○, and ●●●. Many of these activities are presented in a "slate, chalk, and eraser" format that engages all students in answering questions and allows the teacher to quickly assess students' understanding.

Math Messages activate and build on students' prior knowledge and create a context for the material to be learned.

Informing Instruction notes suggest how to use observations of students' work to adapt instruction. These notes are designed to help the teacher anticipate and recognize common errors and misconceptions in students' thinking or to alert the teacher to multiple solution strategies or unique insights students may offer.

Getting Started

Mental Math and Reflexes

Write decimals on the board and ask students to read them. *Suggestions:*

●○○	●●○	●●●
0.5	34.12	0.984
0.76	9.03	0.733
0.14	465.81	0.804

Math Message

Solve Problem 1 on journal page 82.

Study Link 4·2 Follow-Up

Have students share examples of decimals they brought from home. Discuss their meanings and values. Use such language as, *The label on a package of chicken reads "2.89 pounds." 2.89 pounds is between 2 and 3 pounds. It is almost 3 pounds.* Encourage students to continue bringing examples of decimals to display in a Decimals All Around Museum. See the optional ELL Support activity in Part 3 for details.

1 Teaching the Lesson

▶ Math Message Follow-Up

WHOLE-CLASS DISCUSSION

(*Math Journal 1*, p. 82)

Discuss ways to show that 0.3 > 0.15. Be sure to include the following two methods:

▷ Model **decimals** with base-10 blocks. If a flat is ONE, then 0.3 is $\frac{3}{10}$ of the flat, or 3 longs, and 0.15 is $\frac{15}{100}$ of the flat, or 15 cubes. Because 3 longs are more than 15 cubes, 0.3 > 0.15.

▷ Rename one of the decimals so that both decimals have the same number of digits to the right of the decimal point. Do so by appending zeros to the decimal having fewer digits after the decimal point. In this problem, show that 0.3 = 0.30 by trading 3 longs for 30 cubes. Because 30 cubes are more than 15 cubes, 0.30 > 0.15. Therefore, 0.3 > 0.15.

Have students use base-10 blocks to complete Problem 2 on journal page 82.

Ongoing Assessment: Informing Instruction

Watch for students who think 0.3 is less than 0.15 because 3 is less than 15. Modeling the problems with base-10 blocks and then trading longs for cubes can help students understand why zeros can be appended to a decimal without changing its value.

Writing a zero at the end of a decimal corresponds to thinking about the number in terms of the next smaller place. For example, 30 hundredths, 0.30, or 30 cubes is greater than 15 hundredths, 0.15, or 15 cubes. Note how this differs from the situation with whole numbers: With whole numbers, the number with more digits is always greater.

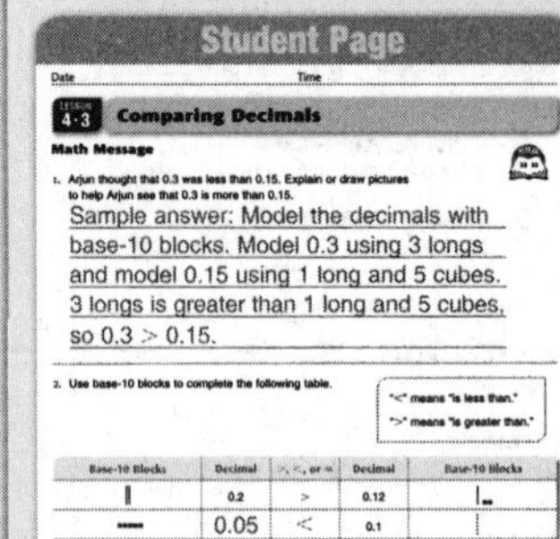

Student Page

Date Time

Lesson 4·3 Comparing Decimals

Math Message

1. Arjun thought that 0.3 was less than 0.15. Explain or draw pictures to help Arjun see that 0.3 is more than 0.15.

Sample answer: Model the decimals with base-10 blocks. Model 0.3 using 3 longs and model 0.15 using 1 long and 5 cubes. 3 longs is greater than 1 long and 5 cubes, so 0.3 > 0.15.

2. Use base-10 blocks to complete the following table.

"<" means "is less than."
">" means "is greater than."

Base-10 Blocks	Decimal	>, <, or =	Decimal	Base-10 Blocks
	0.2	>	0.12	
	0.05	<	0.1	
	0.13	<	0.31	
	0.33	>	0.3	
	1.2	<	2.1	
	0.47	>	0.39	
	2.3	=	2.3	

Math Journal 1, p. 82

Lesson 4·3 251

Recognizing Student Achievement notes highlight specific tasks that can be used for assessment to monitor students' progress toward Grade-Level Goals. The notes identify the expectations for a student who is making adequate progress and point to skills or strategies that some students may be able to demonstrate.

Games played in the classroom, online, and at home provide significant practice in *Everyday Mathematics.* Games are ideal for differentiating instruction as rules and levels of difficulty can be modified easily.

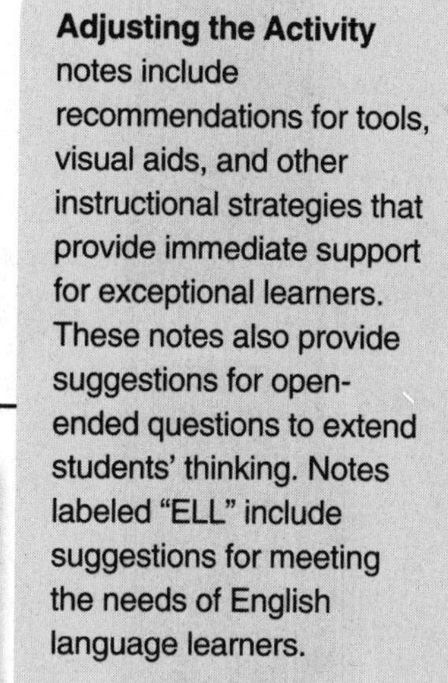
Adjusting the Activity notes include recommendations for tools, visual aids, and other instructional strategies that provide immediate support for exceptional learners. These notes also provide suggestions for open-ended questions to extend students' thinking. Notes labeled "ELL" include suggestions for meeting the needs of English language learners.

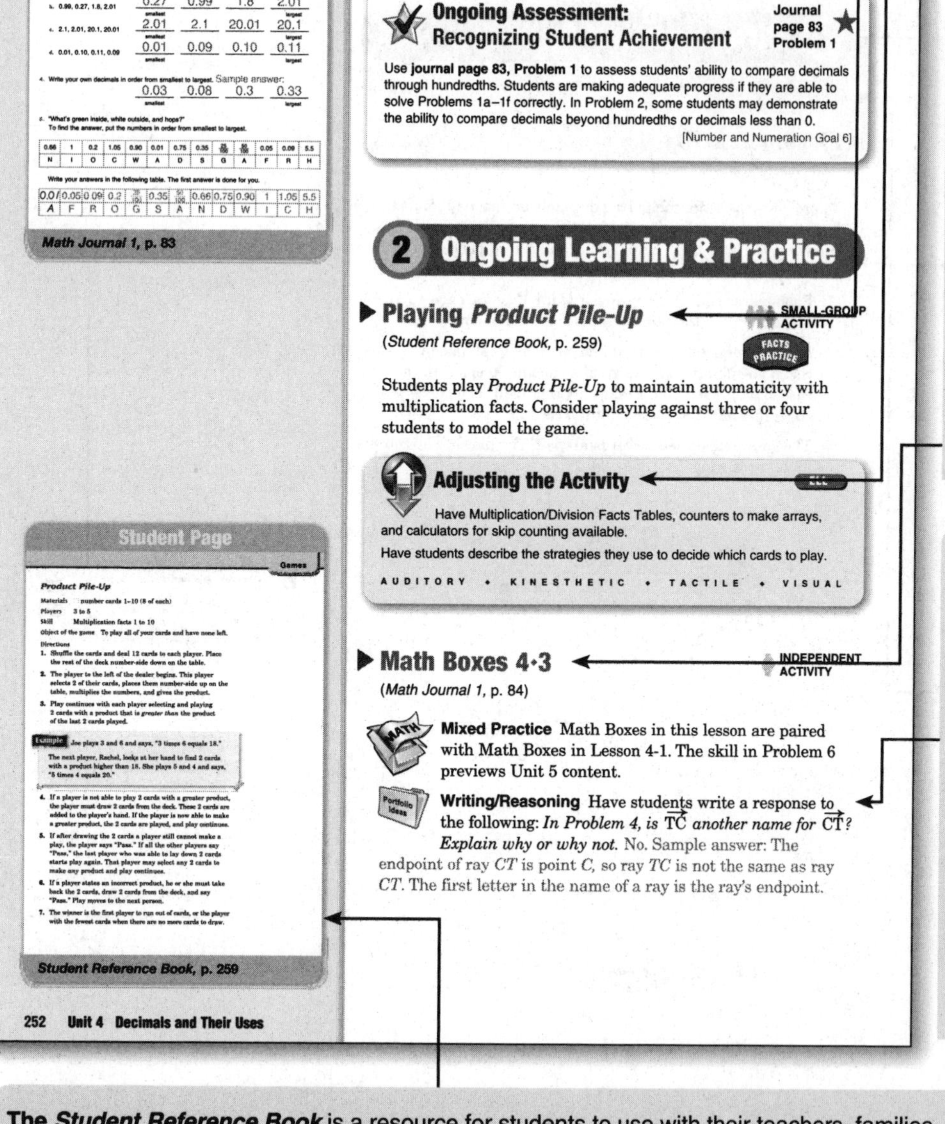
Student Page

Ordering Decimals

Math Journal 1, p. 83

Ordering Decimals

PROBLEM SOLVING · PARTNER ACTIVITY · ELL

(*Math Journal 1*, p. 83)

Students compare and order decimals. Base-10 blocks should be available. English language learners may struggle with understanding the answer to the riddle in Problem 5.

Ongoing Assessment: Recognizing Student Achievement — Journal page 83 Problem 1

Use **journal page 83, Problem 1** to assess students' ability to compare decimals through hundredths. Students are making adequate progress if they are able to solve Problems 1a–1f correctly. In Problem 2, some students may demonstrate the ability to compare decimals beyond hundredths or decimals less than 0.

[Number and Numeration Goal 6]

2 Ongoing Learning & Practice

Playing *Product Pile-Up* — SMALL-GROUP ACTIVITY · FACTS PRACTICE

(*Student Reference Book*, p. 259)

Students play *Product Pile-Up* to maintain automaticity with multiplication facts. Consider playing against three or four students to model the game.

Adjusting the Activity — ELL

Have Multiplication/Division Facts Tables, counters to make arrays, and calculators for skip counting available.

Have students describe the strategies they use to decide which cards to play.

AUDITORY • KINESTHETIC • TACTILE • VISUAL

Math Boxes 4·3 — INDEPENDENT ACTIVITY

(*Math Journal 1*, p. 84)

Mixed Practice Math Boxes in this lesson are paired with Math Boxes in Lesson 4-1. The skill in Problem 6 previews Unit 5 content.

Writing/Reasoning Have students write a response to the following: *In Problem 4, is* $\overrightarrow{TC}$ *another name for* $\overrightarrow{CT}$*? Explain why or why not.* No. Sample answer: The endpoint of ray *CT* is point *C,* so ray *TC* is not the same as ray *CT.* The first letter in the name of a ray is the ray's endpoint.

Student Page

Product Pile-Up

Student Reference Book, p. 259

252 Unit 4 Decimals and Their Uses

Math Boxes are designed to provide distributed practice. Math Boxes routinely revisit recent content to help students build and maintain important concepts and skills. One or two problems on each journal page preview content for the coming unit. Use class performance on these problems as you plan for the coming unit.

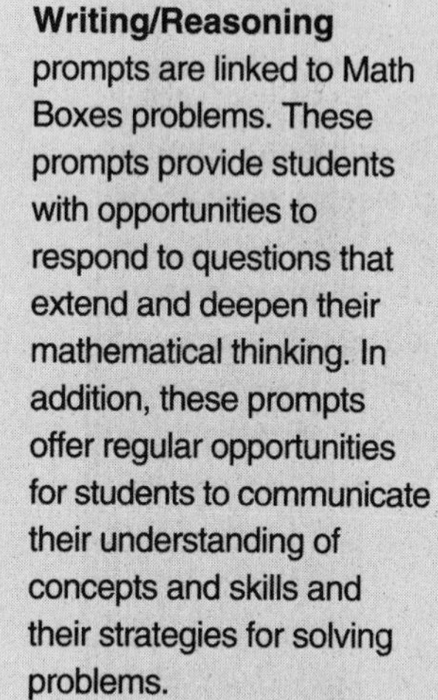
Writing/Reasoning prompts are linked to Math Boxes problems. These prompts provide students with opportunities to respond to questions that extend and deepen their mathematical thinking. In addition, these prompts offer regular opportunities for students to communicate their understanding of concepts and skills and their strategies for solving problems.

The *Student Reference Book* is a resource for students to use with their teachers, families, and classmates. It includes examples of completed problems similar to those students encounter in class, explanations, illustrations, and game directions. The *Student Reference Book* provides excellent support for all students, including English language learners and their families. At Grades 1 and 2, this book is called ***My Reference Book.***

Study Links are the *Everyday Mathematics* version of homework assignments. At Grades 1 through 3, they are called **Home Links.** They can be assigned to promote mathematical discussions at home.

▶ Study Link 4·3

INDEPENDENT ACTIVITY

(*Math Masters,* p. 112)

Home Connection Students order decimals on a number line and find decimals between two given amounts.

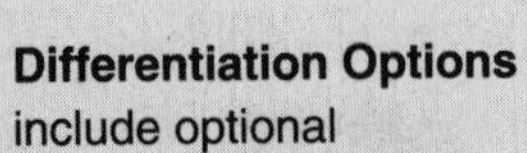

Differentiation Options include optional Readiness, Enrichment, Extra Practice, and ELL Support activities that can be used with individual students, small groups, or the whole class. The activities build on the Key Concepts and Skills highlighted in Part 1 of each lesson. Use them to supplement, modify, or adapt the lesson to meet students' needs.

3 Differentiation Options

Readiness Activities preview lesson content or provide students with alternative strategies for learning concepts and skills.

READINESS

PARTNER ACTIVITY

▶ Playing *Coin Top-It*

5–15 Min

(*Math Masters,* pp. 467 and 506)

To provide experience comparing decimals in a money context, have students play *Coin Top-It.* Ask them to model the amounts shown on the cards with actual coins and record play on *Math Masters,* page 506.

1. Each player cuts apart a copy of *Math Masters,* page 467. Players shuffle the cards and place them facedown.
2. Each player draws one card and says the total amount of the coins. The player with the greater amount keeps both cards. In case of a tie, each player takes another card. The player with the larger amount takes all of the cards.
3. The game ends when no cards are left. The player who collects more cards wins.

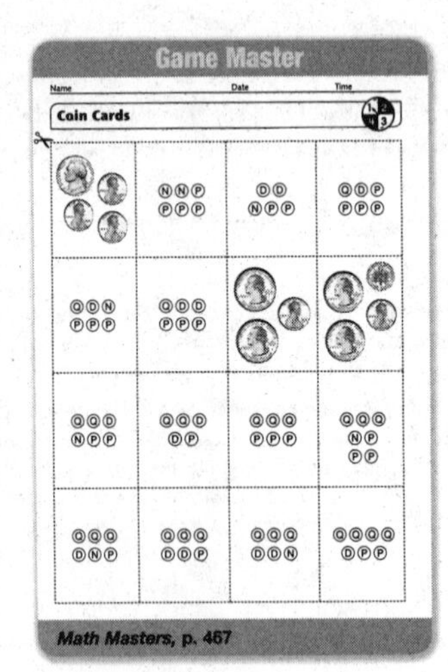

Math Masters, p. 467

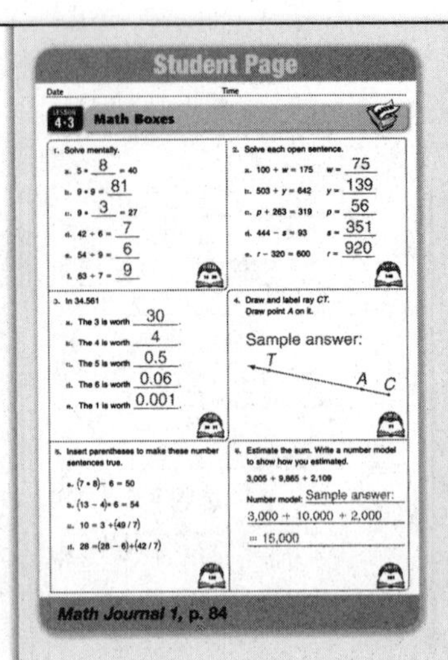

Student Page

Date ____ Time ____

Lesson 4·3 Math Boxes

1. Solve mentally.
   a. $5 * \underline{8} = 40$
   b. $9 * 9 = \underline{81}$
   c. $9 * \underline{3} = 27$
   d. $42 \div 6 = \underline{7}$
   e. $54 \div 9 = \underline{6}$
   f. $63 \div 7 = \underline{9}$
2. Solve each open sentence.
   a. $100 + w = 175$ $w = \underline{75}$
   b. $503 + y = 642$ $y = \underline{139}$
   c. $p + 263 = 319$ $p = \underline{56}$
   d. $444 - s = 93$ $s = \underline{351}$
   e. $r - 320 = 600$ $r = \underline{920}$
3. In 34.561
   a. The 3 is worth 30.
   b. The 4 is worth 4.
   c. The 5 is worth 0.5.
   d. The 6 is worth 0.06.
   e. The 1 is worth 0.001.
4. Draw and label ray *CT.* Draw point *A* on it.
   Sample answer: T A C
5. Insert parentheses to make these number sentences true.
   a. $(7 * 8) - 6 = 50$
   b. $(13 - 4) * 6 = 54$
   c. $10 = 3 + (49 / 7)$
   d. $28 = (28 - 6) + (42 / 7)$
6. Estimate the sum. Write a number model to show how you estimated.
   3,005 + 9,865 + 2,109
   Number model: Sample answer: 3,000 + 10,000 + 2,000 = 15,000

Math Journal 1, p. 84

Study Link Master

Name ____ Date ____ Time ____

Study Link 4·3 Ordering Decimals

Mark the approximate locations of the decimals and fractions on the number lines below. Rename fractions as decimals as necessary.

1. Number line 0.0 to 2 (0.0, 0.25, 0.5, 0.75, 1, 1.25, 1.5, 1.75, 2), marked G A CH D F B E
   A 0.33 *B* 1.6 *C* 0.7 *D* 1.01
   E 1.99 *F* 1.33 *G* 0.1 *H* 0.8
2. Number line 0.0 to 1.2 (0.0, 0.1, 0.2, 0.3, 0.4, 0.5, 0.6, 0.7, 0.8, 0.9, 1.0, 1.1, 1.2), marked JP O LQ I K R M N
   I 0.67 *J* 0.05 *K* $\frac{75}{100}$ *L* 0.49 *M* 0.99
   N 1.15 *O* $\frac{25}{100}$ *P* 0.101 *Q* 0.55 *R* 0.88

Use decimals. Write 3 numbers that are between the following: Sample answers:

3. $5 and $6	$5.05	$5.25	$5.95
4. 4 centimeters and 5 centimeters	4.15 cm	4.5 cm	4.99 cm
5. 21 seconds and 22 seconds	21.4 sec	21.98 sec	21.57 sec
6. 8 dimes and 9 dimes	$0.89	$0.85	$0.82
7. 2.15 meters and 2.17 meters	2.155 m	2.16 m	2.159 m
8. 0.8 meter and 0.9 meter	0.84 m	0.88 m	0.87 m

Practice

9. $x + 17 = 23$ $x = \underline{6}$ 10. $5 * n = 35$ $n = \underline{7}$ 11. $32 / b = 4$ $b = \underline{8}$

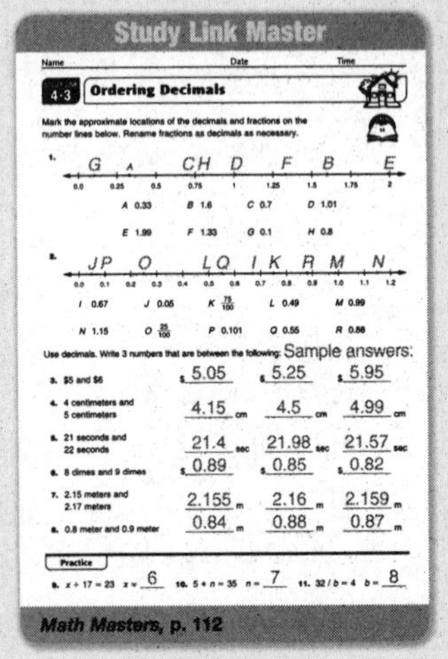

Math Masters, p. 112

Lesson 4·3 253

Enrichment Activities apply or deepen students' understanding of lesson content.

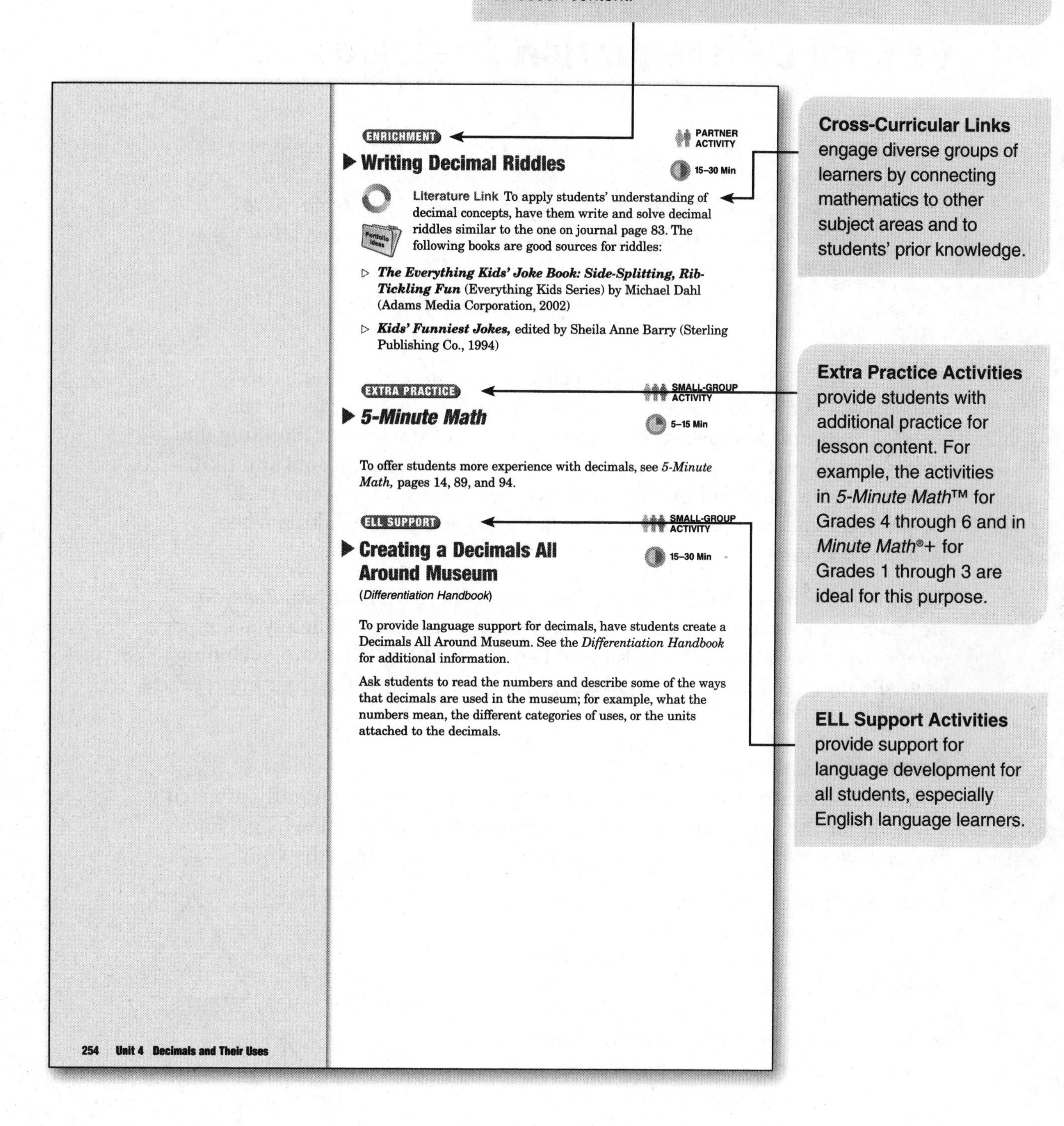

Cross-Curricular Links engage diverse groups of learners by connecting mathematics to other subject areas and to students' prior knowledge.

Extra Practice Activities provide students with additional practice for lesson content. For example, the activities in *5-Minute Math*™ for Grades 4 through 6 and in *Minute Math*®+ for Grades 1 through 3 are ideal for this purpose.

ELL Support Activities provide support for language development for all students, especially English language learners.

Features for Differentiating in *Everyday Mathematics*

General Differentiation Strategies

> *All tasks should respect each learner. Every student deserves work that is focused on the essential knowledge, understanding, and skills targeted for the lesson. Every student should be required to think at a high level and should find his or her work interesting and powerful.*
>
> (Tomlinson 2003, 61, 2: 9)

Each *Everyday Mathematics* lesson focuses on a range of mathematical concepts and skills. The most prominent of these are highlighted in the *Key Concepts and Skills* section at the beginning of the lesson. Planning for differentiated instruction involves analyzing which Key Concepts and Skills are appropriate as learning objectives for individual students and then supporting, emphasizing, and enhancing these concepts and skills when teaching the lesson.

Examples of some of the instructional strategies incorporated into *Everyday Mathematics* lessons are described here. These strategies will help you support, emphasize, and enhance lesson content to ensure that all students, including English language learners, are engaged in the mathematics at their appropriate developmental level.

Framing the Lesson

Lesson introductions set the stage and support learning by mentally preparing students for the content of the lesson or by activating prior knowledge. For example, you might begin a geometry lesson with one of the following:

- Remind students that they were working on 2-dimensional shapes in the last lesson. Have them discuss what they remember about 2-dimensional shapes.

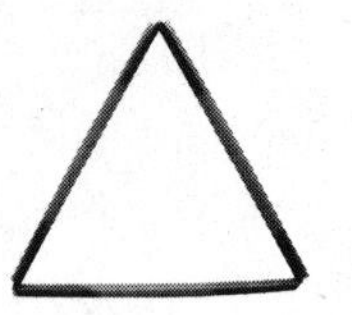

- Tell students that today they are going to build geometric shapes using straws. Ask: *What are some things you know about shapes that will help you with this activity?*

Providing Wait Time

Lessons consist of whole-class, small-group, partner, and independent work. During the whole-class portion of a lesson, allow time for students to think and process information before eliciting answers to questions posed. Waiting even a few seconds for an answer will help many students process information and, in turn, participate more fully in class discussions.

Wait time is also beneficial when you pose Mental Math and Reflexes problems. Encourage students to stop and think before they write on their slates and show their answers. Consider displaying the three steps on a poster. Establish a routine by pointing to the steps in sequence, pausing at each for several seconds.

Establish a routine using Mental Math and Reflexes in which students Think, Write, *and* Show.

Making Connections to Everyday Life

Lessons offer regular opportunities to build on students' everyday life by helping them make connections between their everyday experiences and new mathematics concepts and skills.

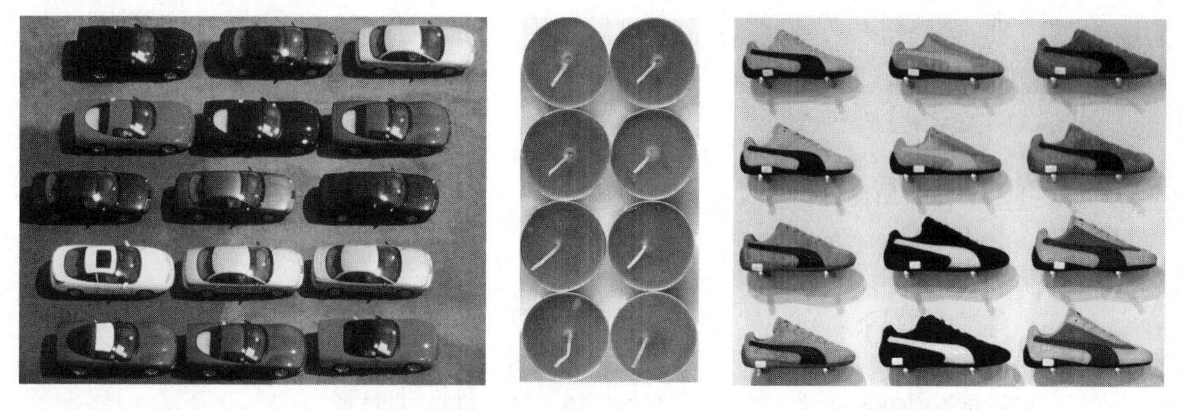

Students build an Array Museum to display examples of arrays found in everyday life. Arrays are closely related to equal-groups situations. If the equal groups are arranged in rows and columns, then a rectangular array is formed. As with equal-group situations, arrays can lead to either multiplication or division problems.

Modeling Concretely

Everyday Mathematics lessons frequently include the use of manipulatives. Make them easily available at all times and for all students. Modeling concretely not only makes lesson content more accessible for some students, but it can also deepen all students' understanding of concepts and skills.

- Have pattern blocks available so students can model fraction computation problems such as $\frac{2}{3} + \frac{1}{6} = \_\_\_$.

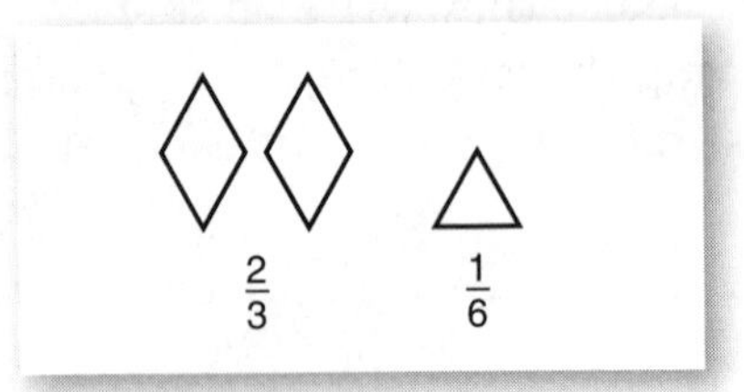

Model the fractions to be added with pattern blocks.

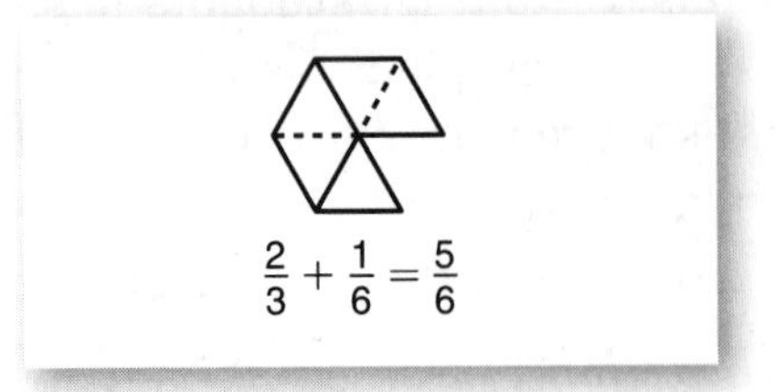

Combine the blocks to show the sum.

- Have stick-on notes available for making line plots and finding the median.

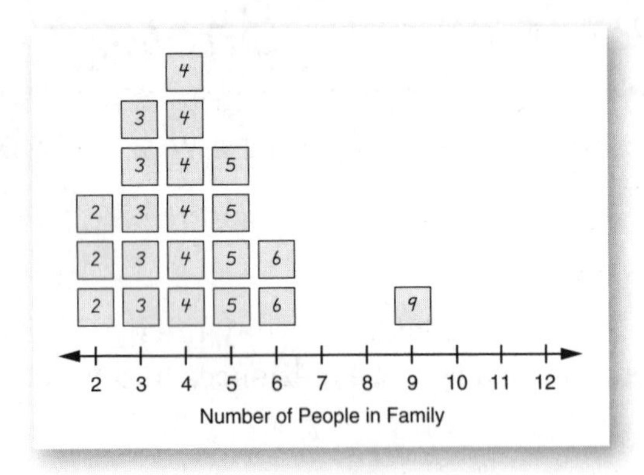

Use stick-on notes to make a line plot showing the number of people students have in their families.

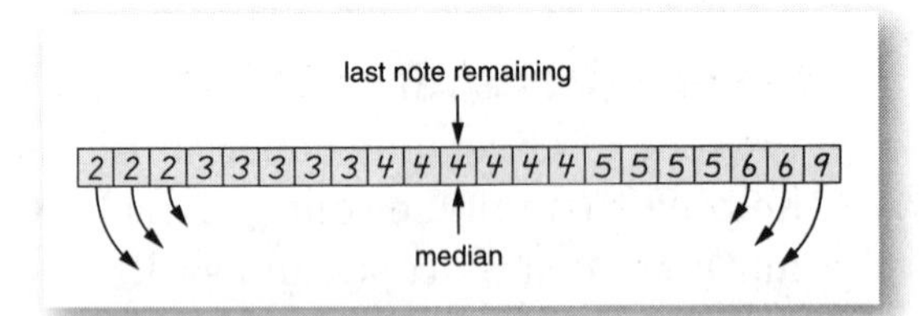

Find the median by lining up the stick-on notes and removing notes two at a time, one from each end, until only one or two notes are left.

- Have base-10 blocks available so students can model place value, addition, and subtraction.

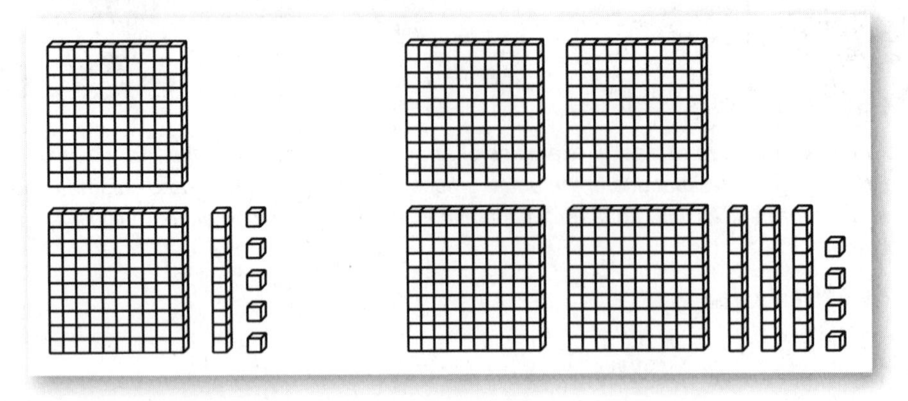

215 + 434 = ______

Modeling Visually

Classrooms tend to be highly verbal places, and this can be overwhelming for some students. Simple chalkboard drawings, diagrams, and other visual representations can help students make sense of the flow of words around them and can also help them connect words to the actual items.

- Use pictures to model even and odd numbers.

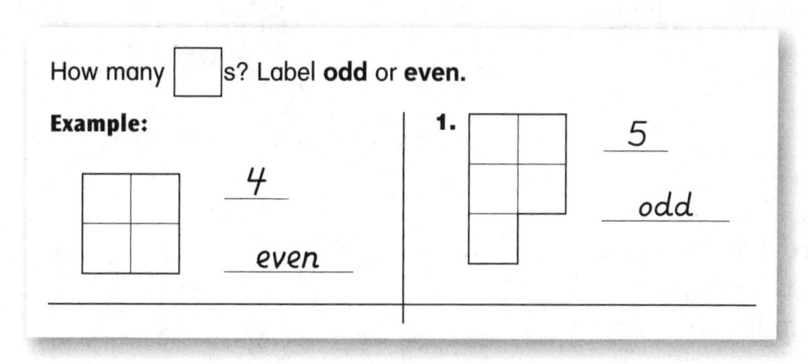

- Use arrays to model square numbers.

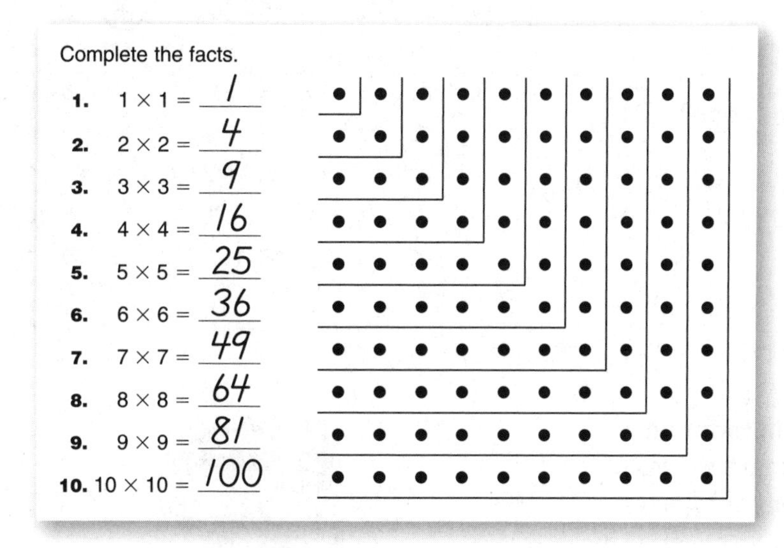

- Use a number line to visually model division by a fraction.

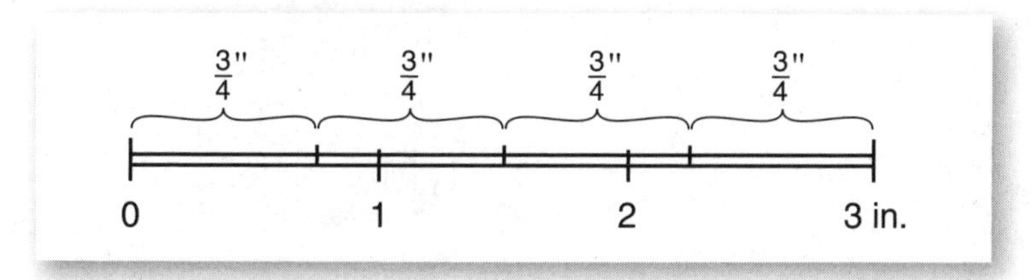

To illustrate division of a whole number by a fraction, students partition a 3-inch segment into equal $\frac{3}{4}$-inch segments. Students ask the question, "How many $\frac{3}{4}$-inch segments are in 3 inches?" They answer the question by counting the number of line segments; in this case, there are 4 equal segments. Students then write the number sentence, $3 \div \frac{3}{4} = 4$.

Modeling Physically

Lessons suggest ways to have students demonstrate concepts and skills with gestures or movements. This strategy helps many students better understand and retain the concept or skill.

- Have students model the concept of *parallel* by holding their arms in front of them, parallel to each other.

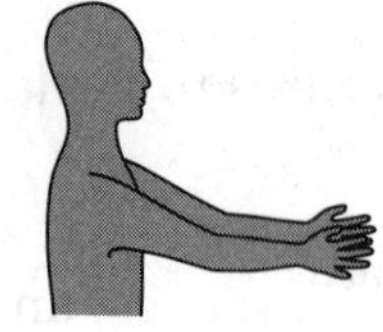

A physical model for parallel line segments

- Have students model addition and subtraction problems by moving their fingers on number lines or number grids. A number-grid master can be found on page 148 of this handbook.

–9	–8	–7	–6	–5	–4	–3	–2	–1	0
1	2	3	4	5	6	7	8	9	10
11	12	13	14	15	16	17	18	19	20
21	22	23	24	25	26	27	28	29	30
31	32	33	34	35	36	37	38	39	40
41	42	43	44	45	46	47	48	49	50
51	52	53	54	55	56	57	58	59	60
61	62	63	64	65	66	67	68	69	70
71	72	73	74	75	76	77	78	79	80
81	82	83	84	85	86	87	88	89	90
91	92	93	94	95	96	97	98	99	100
101	102	103	104	105	106	107	108	109	110

$15 + 33 =$ ______

- Have students skip count on a calculator while doing a class count. This strategy reinforces counting visually by showing the numbers while at the same time physically engaging students.

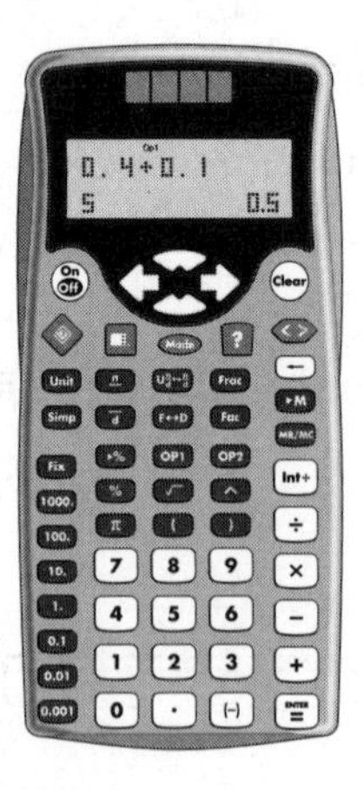

Program a TI-15 calculator to count by tenths. Clear the calculator. Enter Op1 + *0.1* Op1 *0* Op1 *and repeatedly enter* Op1 *without clearing the calculator.*

Providing Organizational Tools

Lessons provide a variety of tools to help students organize their thinking. Have students use diagrams, tables, charts, and graphs when these materials are included in lessons and as appropriate. This is another way to make the lesson content more accessible for some students while at the same time deepening other students' understanding of concepts and skills.

- Have students use Venn diagrams to compare and contrast properties of numbers, shapes, and so on.

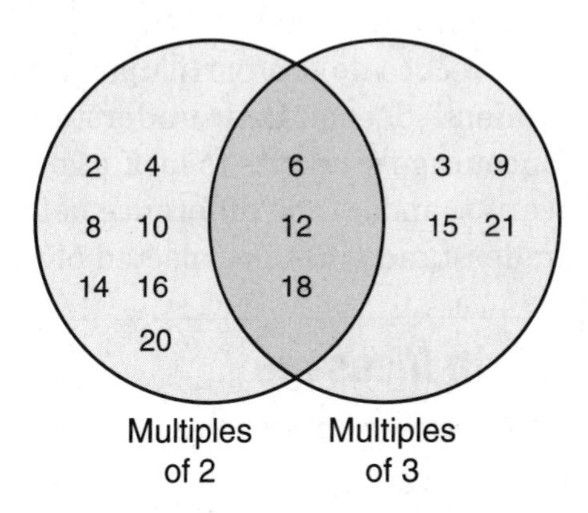

Blank masters for Venn diagrams can be found on pages 149 and 150 of this handbook.

- Have students use situation diagrams to model operations. Blank masters for these diagrams can be found on pages 151 and 152 of this handbook.

rows	chairs per row	chairs in all
3	?	15

Adriana set up chairs in her backyard for a play. She had 15 chairs in all. She made 3 rows. How many chairs were in each row?

Total ?	
Part 47	Part 15

Malcolm had 47 pennies in a jar in his room. His brother had 15 pennies. How many pennies did they have in all?

- Provide students with place-value charts or have them draw their own. Have them write numbers in the charts as dictated, for example, the number that has a 3 in the thousands place, a 2 in the ones place, a 4 in the ten-thousands place, and a 0 everywhere else.

Ten Thousands	Thousands	Hundreds	Tens	Ones
4	3	0	0	2

Engaging Students in Talking about Math

Lessons often suggest discussion prompts or questions that support the development of good communication skills in the context of mathematics. Although finding the correct solution is one important goal, *Everyday Mathematics* lessons also emphasize sharing and comparing solution strategies. This type of "math talk" involves not only explaining what is done (explanation), but also why it is done (reasoning), and why it is correct or incorrect to do it a particular way (justification). These discussions help students deepen their understanding of mathematical concepts and processes. Encourage students to look at other students when they are speaking. You may want to model the difference between hearing and listening to help students understand what is expected of them.

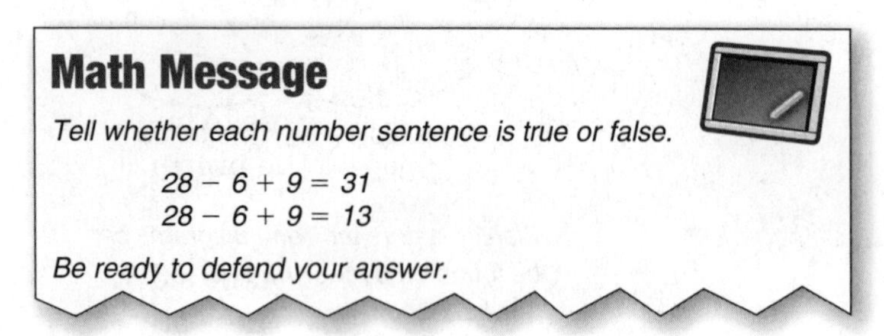

Some students may work the problems from left to right and determine that the first number sentence is true. Other students may decide that the second number sentence is true by first adding 6 and 9 and then subtracting the sum from 28. Others may reason that both sentences could be true, depending on what you do first. This Math Message problem and the resulting discussion serve as an introduction to the use of parentheses in number sentences that involve more than one operation.

Engaging Students in Writing about Math

Journal pages and assessment problems frequently prompt students to explain their thinking and strategies in words, pictures, and diagrams. Writing offers students opportunities to reflect on their thinking and can help you assess their mathematical understandings and communication skills. Exit Slips and Math Logs are ideal places for students to record their thinking.

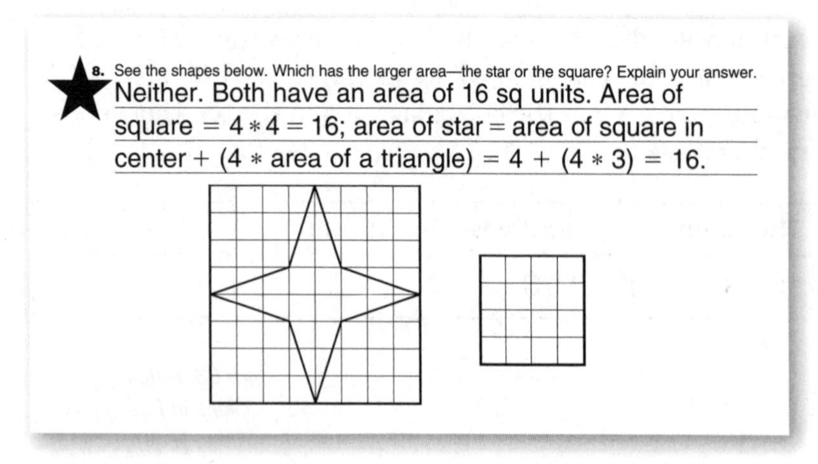

Students demonstrate their understanding of area by responding to the question on the journal page.

Using Key Concepts and Skills

Each *Everyday Mathematics* lesson provides students with opportunities to explore a variety of mathematics. This variety allows you to target appropriate concepts and skills for individual students.

Shown below are the Key Concepts and Skills in Lesson 3-4 of *Third Grade Everyday Mathematics.*

Key Concepts and Skills
- Use basic facts to find perimeter. [Operations and Computation Goal 1]
- Model polygons with straws; identify and describe polygons. [Geometry Goal 2]
- Measure sides of polygons to the nearest inch. [Measurement and Reference Frames Goal 1]
- Add side lengths to find perimeter. [Measurement and Reference Frames Goal 2]

At the beginning of the lesson, students use straws and connectors to build the polygons in Problems 1 and 2 on journal page 63 and compare the properties of these polygons. Students then work on the journal page.

- Modeling and describing polygons and measuring the lengths of the sides may be reasonable skills to target for some students in this lesson. These students might complete only Problems 1 and 2 on journal page 63.

- Problems 1 and 2 may be the most important ones for some students to complete. Encourage students to complete these problems first and to finish the remainder of the page if they have time, comparing their answers with one another. Circulate and assist.

- Finding perimeters may be a reasonable skill to emphasize for some students. If students completed Problem 4 by drawing a rectangle with a perimeter of 20, ask them to apply their understanding of perimeter by drawing other rectangles on a grid with perimeters of 20. Have students write an explanation of how they can be sure they found all such rectangles.

- Some students may need more time to complete all the problems. Have students who do not complete the page during the course of the lesson complete it later as time and experience allow.

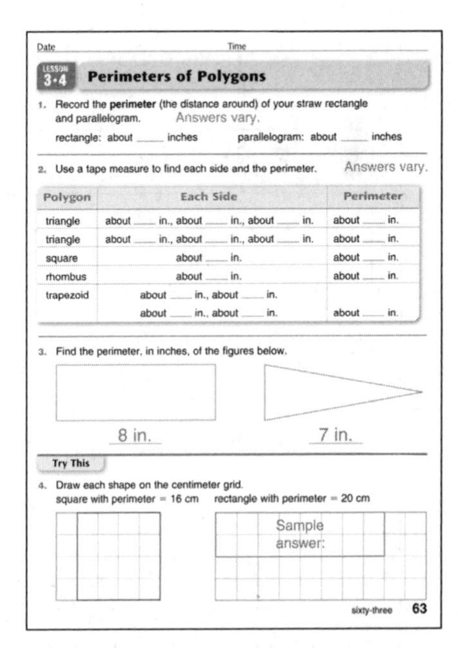

Date Time

LESSON 3·4 **Perimeters of Polygons**

1. Record the **perimeter** (the distance around) of your straw rectangle and parallelogram. Answers vary.

rectangle: about ____ inches parallelogram: about ____ inches

2. Use a tape measure to find each side and the perimeter. Answers vary.

Polygon	Each Side	Perimeter
triangle	about ____ in., about ____ in., about ____ in.	about ____ in.
triangle	about ____ in., about ____ in., about ____ in.	about ____ in.
square	about ____ in.	about ____ in.
rhombus	about ____ in.	about ____ in.
trapezoid	about ____ in., about ____ in. about ____ in., about ____ in.	about ____ in.

3. Find the perimeter, in inches, of the figures below.

8 in. 7 in.

Try This

4. Draw each shape on the centimeter grid.
square with perimeter = 16 cm rectangle with perimeter = 20 cm

Sample answer:

sixty-three 63

Math Journal 1, *page 63, reflects the Key Concepts and Skills in Lesson 3-4 of* Third Grade Everyday Mathematics.

Summarizing the Lesson

Lesson summaries offer students a chance to bring closure to the lesson, reflect on the concepts and skills they have learned, and pose questions they may still have about the lesson content. Exit Slips and Math Logs are ideal places for students to record their reflections about what they learned. For example, a lesson on measurement might close with one of the following:

- Have students describe what they learned about standard units of linear measure.

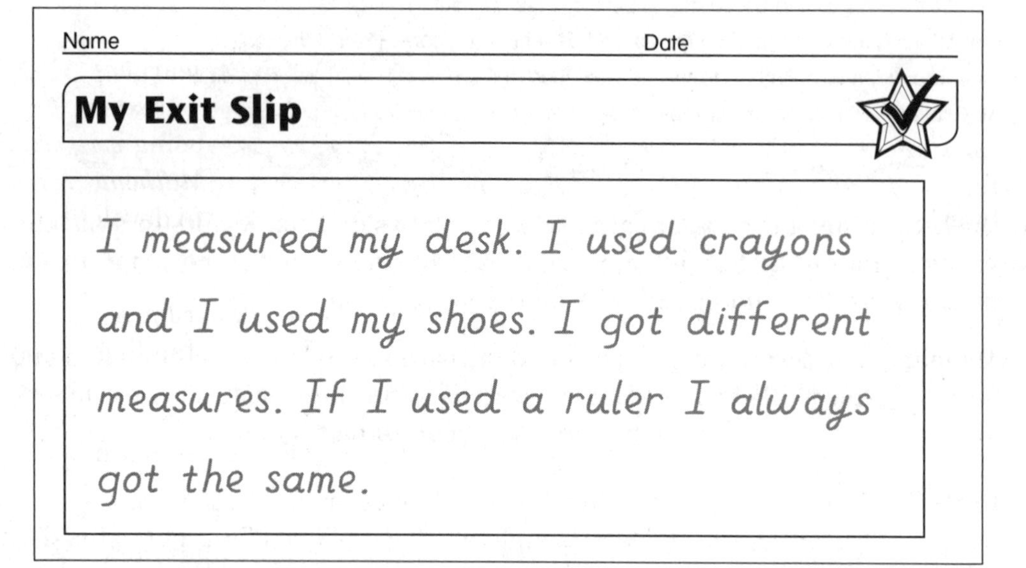

Name Date

My Exit Slip

I measured my desk. I used crayons and I used my shoes. I got different measures. If I used a ruler I always got the same.

Using an Exit Slip, a student describes what she learned.

- Have students record what they know about using a ruler to measure length.

My Exit Slip

When I use a ruler to measure length I have to line up one edge of the line segment with the 0 on the ruler. Sometimes the 0 isn't at the edge.

Using an Exit Slip, a student explains how he uses a ruler.

Vocabulary Development

> *It is of interest to note that while some dolphins are reported to have learned English—up to fifty words in correct context—no human being has been reported to have learned dolphinese.*
>
> —Carl Sagan
> (Robertson 1998, 364)

The most effective way for students, including English language learners, to learn new words is to encounter them repeatedly in meaningful contexts. When the meaning of a new word is understood, real mastery requires using it in conversation and writing. With this principle in mind, *Everyday Mathematics* incorporates many opportunities within the lessons for students to develop vocabulary. *For example:*

- Topics and concepts are regularly revisited throughout the program, so students are constantly building on and deepening their understanding of mathematical terms from previous lessons.
- Hands-on, interactive, and visual activities in each lesson ensure that new words are introduced in clear, comprehensible ways.
- Sharing solutions and explanations, along with cooperative group work, ensures that students have opportunities to use new vocabulary purposefully.

Examples of helpful strategies are described here.

Providing Visual References

Suggest visual references to provide support for the use and development of mathematical language.

- Have students underline the names of the pattern blocks with a pencil that is the same color as the corresponding block to help students associate the words with the shapes.

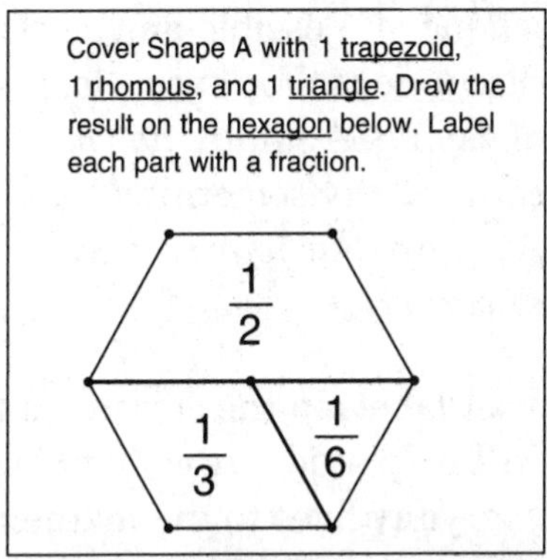

- Record the words and the number model for a number story.

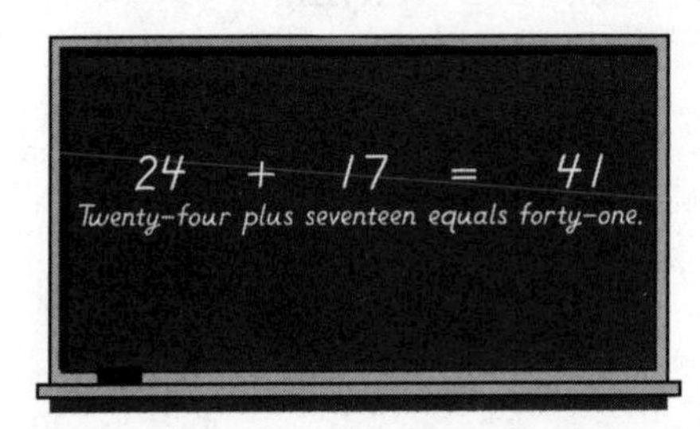

Using the Student Reference Book

The *Student Reference Book* is a rich resource for definitions and examples of vocabulary. Teach students how to use the table of contents, index, and glossary to make optimal use of this resource. The *Student Reference Book* is also a good source of illustrations that English language learners often find useful.

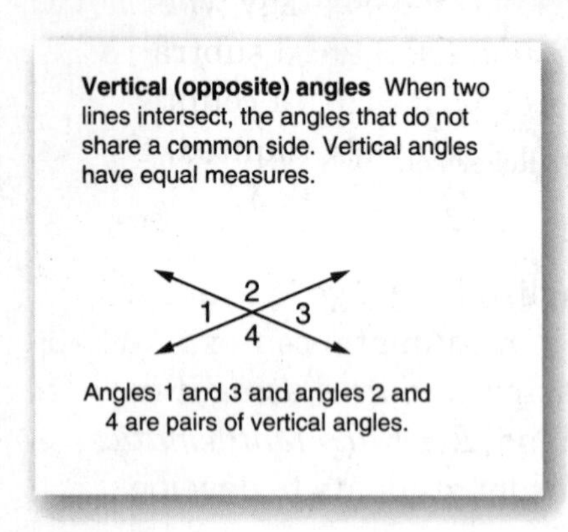

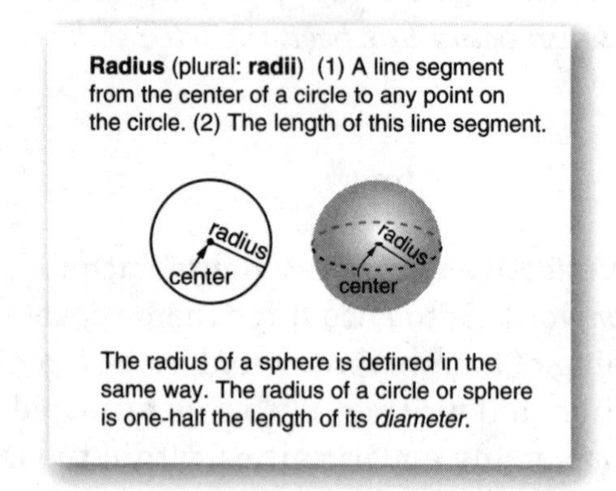

Creating a Language-Rich Environment

Support students' development of their mathematics vocabulary by immersing them in a language-rich environment. Seeing, hearing, and using new terms in meaningful ways will help them navigate through the language-rich mathematics lessons and will support their development of stronger communication skills.

- Display new vocabulary on a Math Word Wall. Include illustrations so that students can make sense of the words and use them in their speech and writing.

- Use mathematical terminology whenever possible during class discussions. For example, instead of saying, *A square has four corners,* say something like, *A square has four vertices, or corners.*

- Post labels in the classroom that will help students connect their everyday lives to the mathematics they are studying.

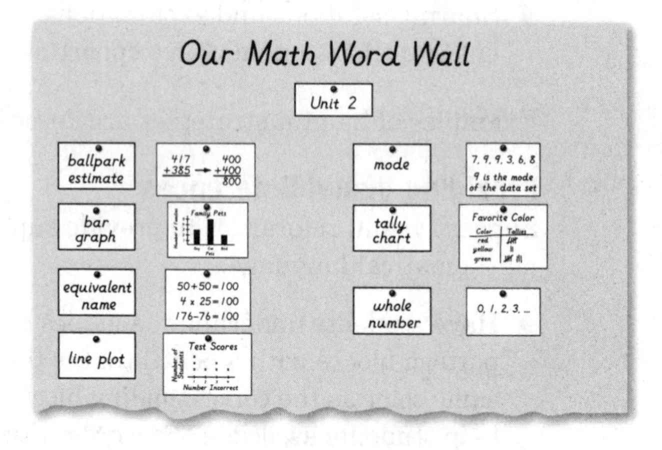

Samples from a Math Word Wall

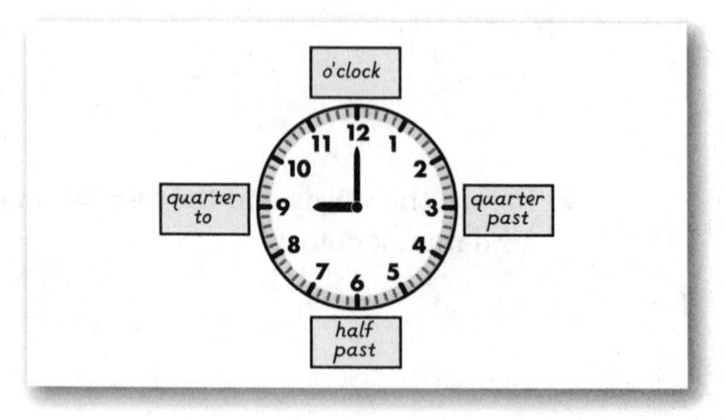

Label your classroom clock.

Recording Key Ideas

On the board or Class Data Pad, record in words, symbols, and pictures the key ideas or key solution steps that students share during class discussions.
For example:

During a class discussion, one student shares the following strategy when explaining how he solved 31 – 14. "First I took 10 from the 14 and subtracted it from 30 because there is a 30 and a 1 in 31. I had 21 left. Then I counted back 4 more for the 4 I had left from 14. I started with 21, then counted, 20, 19, 18, 17. I was holding up four fingers, so I knew I counted back 4."

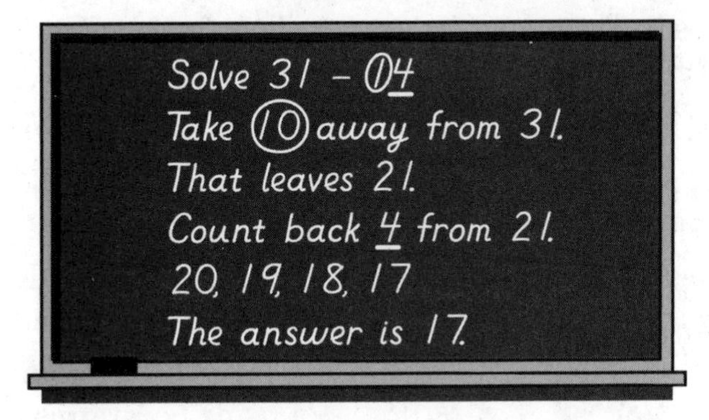

Record the steps of a student's strategy on the board.

Clarifying the Meaning of Words

Lessons routinely highlight potentially confusing words and provide suggestions for clarifying their meanings. Such words include those with multiple meanings, such as *power,* and homophones (words that are pronounced the same, but differ in meaning), such as *sum* and *some.*

- Discuss the everyday versus the mathematical usage of the word *change.*
- Discuss the different meanings of the word *power* in the terms *fact power* and *power of a number.*
- Write *some* and *sum* on the board. Discuss and clarify the meaning of each word.

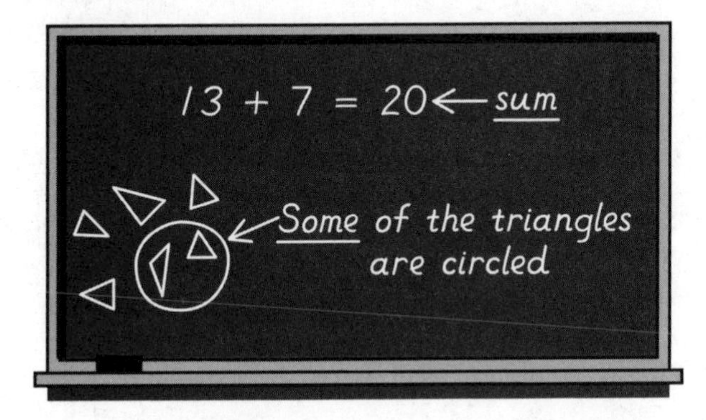

Illustrate the difference between sum *and* some.

Games

Frequent practice is necessary for students to build and maintain strong mental-arithmetic skills and reflexes. There are many opportunities in *Everyday Mathematics* for practice through games. Games are not merely attractive add-ons but an essential component of the *Everyday Mathematics* program and curriculum.

Everyday Mathematics games are important for these reasons:

- Games help students develop the ability to think critically and solve problems. The variety of games in *Everyday Mathematics* lays the foundation for increasingly difficult concepts and helps students develop sophisticated solution strategies.

- Games provide an effective and interactive way for students to practice and master basic concepts and skills. Practice through playing games not only builds fact and operation skills, but often reinforces other concepts and skills, such as calculator use, money exchange, geometric intuition, and ideas about probability.

- Games have advantages over paper-and-pencil drills.

Games	Paper-and-Pencil Drills
Present enjoyable ways to practice skills	Tend toward tedium and monotony
Can be played during free time, lunch and recess, or even at home	Are used only during required class time
Are worksheet-free	Are worksheet-based
Are easily adaptable for a class of students who need to practice a wide range of skills at a variety of levels	Require a variety of worksheets to practice different skills at a variety of levels
Provide immediate insight into students' understanding through their discussions and conversations about mathematics	Result in attempts to understand students' thinking while grading worksheets that are days old

Spend some time learning the *Everyday Mathematics* games so that you understand how much they contribute to students' mathematical progress and can join in the fun.

Using Games in the Classroom

Games can be used in many ways. Consider these ideas for making games both enjoyable and educational for all students:

- Establish a routine to provide all students the opportunity to play games at least two or three times each week for a total of about one hour per week. Practice is most effective when it is distributed, so several short practice sessions are preferable to one large block of time.

- Establish a routine for playing games as a regular part of your math class rather than as a reward for completing assigned work. It is important that all students have time to play games, especially students who work at a slower pace or who may need more practice than their classmates do. This way, students who need the practice the most will not miss out.

- Set up a Games Corner with some of the students' favorite games. Be sure to include all of the gameboards, materials, and game record sheets needed. Consider creating a task card for each game. Encourage students to visit this corner during free time. Change the games menu frequently to correspond with concepts currently taught in your classroom and to offer students additional practice and review of particular skills.

> *LANDMARK SHARK*
> *See Student Reference Book*
> *pages 325 and 326 for instructions.*
> ***Materials you need:***
> *Landmark Shark score sheet*
> *(Math Masters, page 457)*
> *Landmark Shark cards*
> *(Math Masters, page 456)*
> *Everything Math Deck*

Sample task card for Landmark Shark

- Establish game stations where students can rotate to a new station about every 15 minutes. Station time can occur at the beginning or the end of a lesson, during the entire mathematics time, or when a substitute teacher is in the classroom. Consider asking parent volunteers to assist at stations. Provide parents with game directions ahead of time so they are familiar with the rules and with the concepts or skills practiced.

- Monitor students when they play games. Ask students to explain the concept or skill they are practicing or describe strategies they are using.

- Consider students' strengths carefully when pairing or grouping them. Group students so that they can support one another's learning.
  - If it is a new game, consider pairing students who will readily understand and implement the rules with students who may need assistance learning the game.
  - If a familiar game can be played at a variety of levels, consider pairing students who are working at the same level.
- Have students complete game record sheets so they are accountable for the work they do. Alternatively, have students complete Exit Slips summarizing the concepts or skills they practiced.

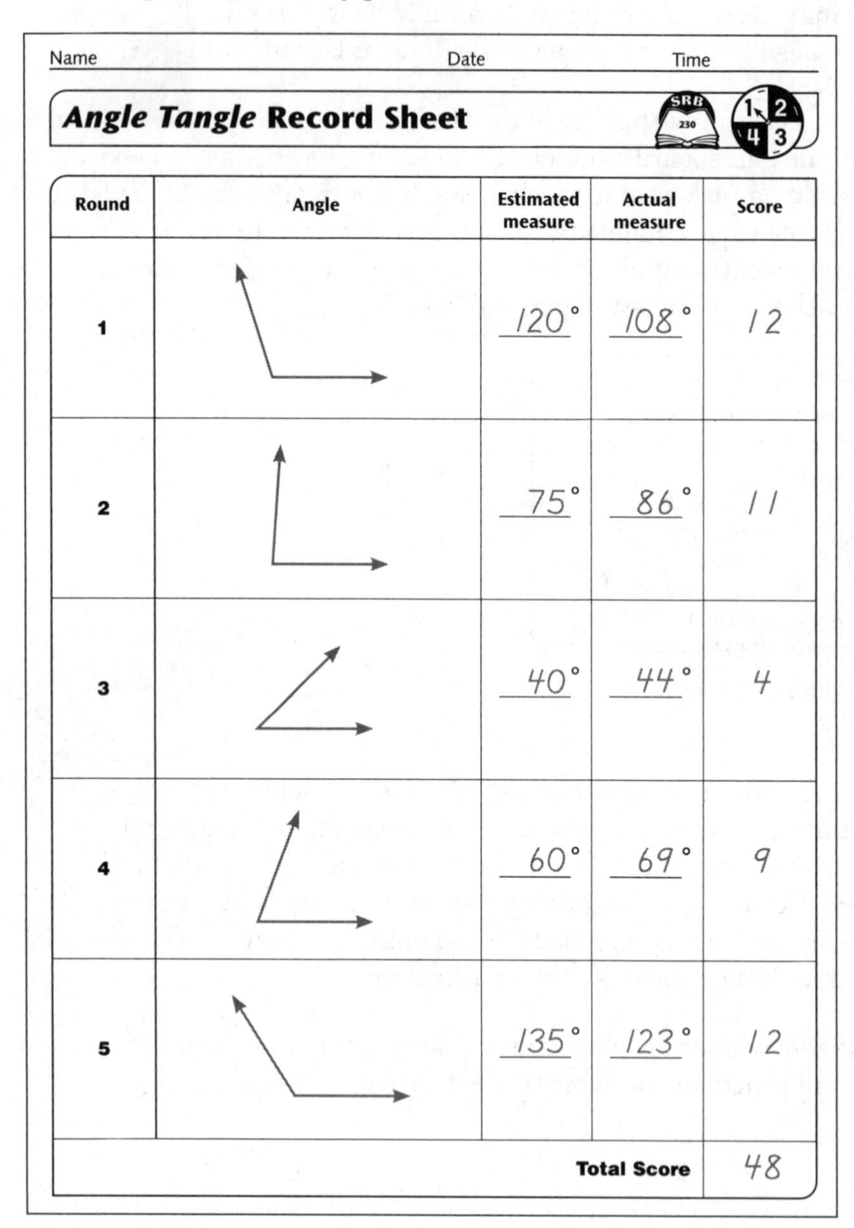

Name Date Time

***Angle Tangle* Record Sheet**

SRB 230

Round	Angle	Estimated measure	Actual measure	Score
1		120°	108°	12
2		75°	86°	11
3		40°	44°	4
4		60°	69°	9
5		135°	123°	12
			Total Score	48

Fourth- through sixth-grade students complete an Angle Tangle *record sheet.*

Modifying Games

Games are easily adapted to meet a variety of practice needs. For example, you can engage all students in the same game at a variety of levels. The modification strategies suggested below can be used for most games included in *Everyday Mathematics*. For specific variations, see the game adaptations in the unit-specific section of this handbook beginning on page 49.

- Modify the level of difficulty of games by targeting a certain range of numbers for students working at different levels. Because numbers in most games are generated randomly, you can modify blank spinners, decrease or increase the number of dice, roll polyhedral dice, or use specific sets of number cards.

Two 6-sided dice for regular game play

Two 8-sided dice to increase the range of numbers

One 10-sided die (0 through 9) to decrease the range of numbers

- Modify the level of difficulty of games by encouraging students to play a mental-math version of a game in which students would normally use paper and pencil to calculate scores.

- See whether variations of a game are available so you can target different concepts or skills or different levels for students appropriately. Many *Everyday Mathematics* games provide a range of practice options by including a variety of gameboards or rules.

Hitting Table	
1 to 10 Facts	
1 to 21	Out
24 to 45	Single (1 base)
48 to 70	Double (2 bases)
72 to 81	Triple (3 bases)
90 to 100	Home Run (4 bases)

Hitting Table	
10 * 10s Game	
100 to 2,000	Out
2,100 to 4,000	Single (1 base)
4,200 to 5,400	Double (2 bases)
5,600 to 6,400	Triple (3 bases)
7,200 to 8,100	Home Run (4 bases)

Versions of Baseball Multiplication *provide practice with different levels of multiplication skills, for example, facts for 1 through 10 and extended facts.*

- Have slates on hand for students to draw pictures as they work through problems.

- Make various manipulatives, such as coins and bills, base-10 blocks, and counters available to provide concrete models for practicing concepts and skills.

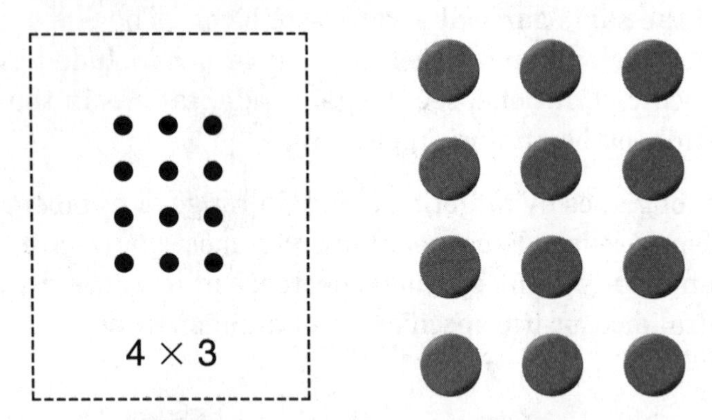

To reinforce multiplication concepts, have students use counters to build the array shown on an Array Bingo *card.*

- Introduce game-specific or mathematical vocabulary with visual cues, such as writing the terms on the board, as well as auditory support, such as having the class repeat the word aloud as a group. Use new vocabulary consistently and be careful to avoid interchanging or substituting synonymous terms, which can cause confusion for some students.

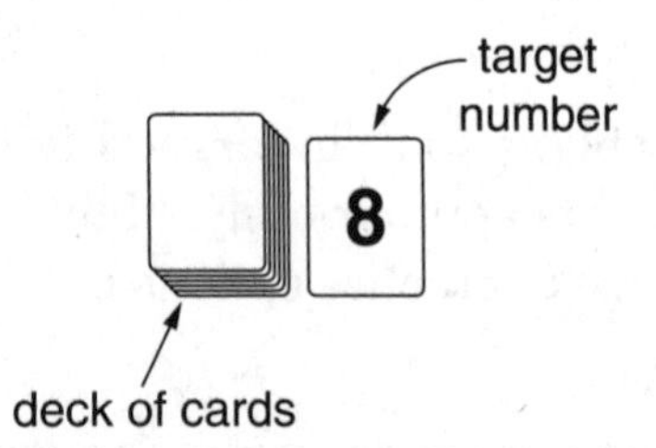

Target number *is a term frequently used in game play.*

- Modify the difficulty of games involving target numbers by limiting the numbers that students use. As students gain proficiency, provide larger numbers. Try this with decimals also.

Students playing Hit the Target *agree on a 2-digit multiple of 10 as the target number for each round. Players then select a starting number and use their calculators to add or subtract to change the starting number to the target number. You can modify the game by suggesting that students choose as the target number a multiple of 10 less than or equal to 40 or a 3- or 4-digit multiple of 100.*

Name ______ Date ______ Time ______

***Hit the Target* Record Sheet**

Round 1

Target number: 30

Starting Number	Change	Result	Change	Result	Change	Result
2	+10	12	+20	32	−2	30

Round 2

Target number: 1,100

Starting Number	Change	Result	Change	Result	Change	Result
317	+700	1,017	+100	1,117	−17	1,100

Round 3

Target number: ______

Starting Number	Change	Result	Change	Result	Change	Result

Round 4

Target number: ______

Starting Number	Change	Result	Change	Result	Change	Result

457

◆ Provide tools such as Addition/Subtraction or Multiplication/Division Facts Tables or calculators for students to check facts quickly and assist them in playing games that require fluency with facts they are learning.

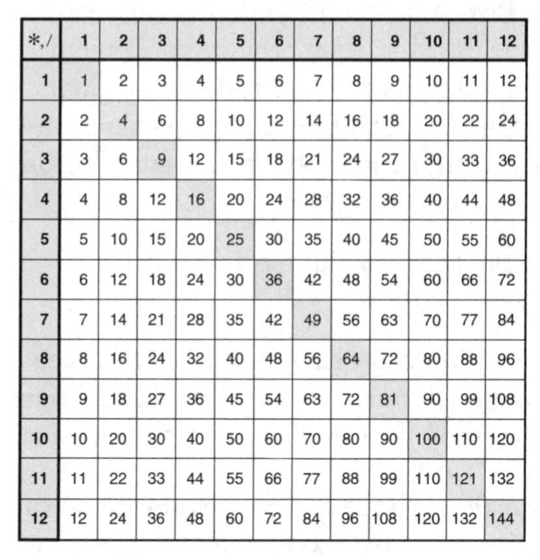

*,/	1	2	3	4	5	6	7	8	9	10	11	12
1	1	2	3	4	5	6	7	8	9	10	11	12
2	2	4	6	8	10	12	14	16	18	20	22	24
3	3	6	9	12	15	18	21	24	27	30	33	36
4	4	8	12	16	20	24	28	32	36	40	44	48
5	5	10	15	20	25	30	35	40	45	50	55	60
6	6	12	18	24	30	36	42	48	54	60	66	72
7	7	14	21	28	35	42	49	56	63	70	77	84
8	8	16	24	32	40	48	56	64	72	80	88	96
9	9	18	27	36	45	54	63	72	81	90	99	108
10	10	20	30	40	50	60	70	80	90	100	110	120
11	11	22	33	44	55	66	77	88	99	110	121	132
12	12	24	36	48	60	72	84	96	108	120	132	144

Multiplication/Division Facts Table

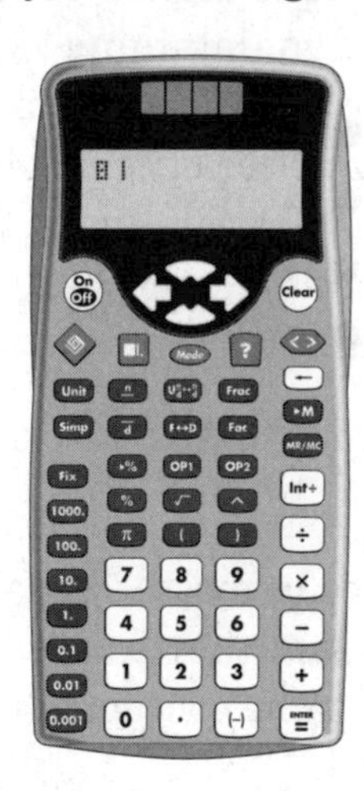

◆ Use illustrations to depict game directions. Create illustrations before introducing the game or during class discussion while introducing the game. Alternatively, have students create the illustrations after they have played the game. Students can refer to the illustrated instructions each time the game is revisited.

◆ Encourage questions and discussion during games so students can use new vocabulary.

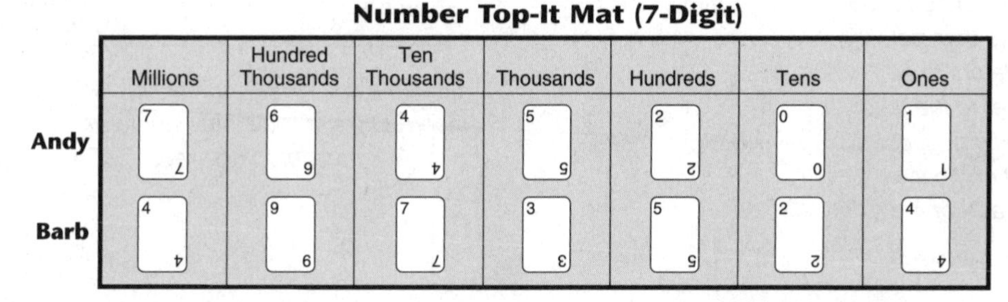

Number Top-It Mat (7-Digit)

	Millions	Hundred Thousands	Ten Thousands	Thousands	Hundreds	Tens	Ones
Andy	7	6	4	5	2	0	1
Barb	4	9	7	3	5	2	4

Students playing Number Top-It *use randomly generated digits to build the largest number possible. Encourage students to discuss and compare strategies for deciding where to place the digits.*

Math Boxes

In *Everyday Mathematics,* Math Boxes are one of the main components for reviewing and maintaining skills. Math Boxes are not intended to reinforce the content of the lesson in which they appear. Rather, they provide continuous distributed practice of concepts and skills targeted in the Grade-Level Goals. It is not necessary for students to complete the Math Boxes page on the same day the lesson is taught, but it is important that the problems for each lesson are completed.

Several features of Math Boxes pages make them useful for differentiating instruction:

- Math Boxes in most lessons are linked with Math Boxes in one or two other lessons so that they have similar problems. Because linked Math Boxes pages target the same concepts and skills, they may be useful as extra practice tools.
- Writing/Reasoning prompts in the *Teacher's Lesson Guide* provide students with opportunities to respond to questions that extend and deepen their mathematical thinking. Using these prompts, students communicate their understanding of concepts and skills and their strategies for solving problems.
- SRB 55–57 Many Math Boxes problems include an icon for the *Student Reference Book*. This cue tells students where they can find help for completing the problems.
- One or two problems on each Math Boxes page preview content from the coming unit. Use these problems identified in the *Teacher's Lesson Guide* to assess student performance and to build your differentiation plan.
- The multiple-choice format of some problems provides students with an opportunity to answer questions in a standardized-test format. The choices include *distractors* that represent common student errors. Use the incorrect answers to identify and address students' needs.

3. Adena drew a line segment $\frac{3}{4}$ inch long. Then she erased $\frac{1}{2}$ inch. How long is the line segment now? Fill in the circle next to the best answer.

Ⓐ $\frac{4}{6}$ in.

Ⓑ $\frac{2}{2}$ in.

Ⓒ $\frac{1}{4}$ in.

Ⓓ $1\frac{1}{4}$ in.

Students choosing $\frac{2}{2}$ in. may have incorrectly subtracted the numerators and denominators. Incorrect algorithm: $\frac{3}{4} - \frac{1}{2} = \frac{3-1}{4-2} = \frac{2}{2}$

Using Math Boxes in the Classroom

Math Boxes can be used in many ways. Consider these ideas for making Math Boxes a productive learning experience for all students:

- ◆ Create a cardstock template that allows students to focus on only one problem at a time. Or, have students use stick-on notes to cover all but one problem.

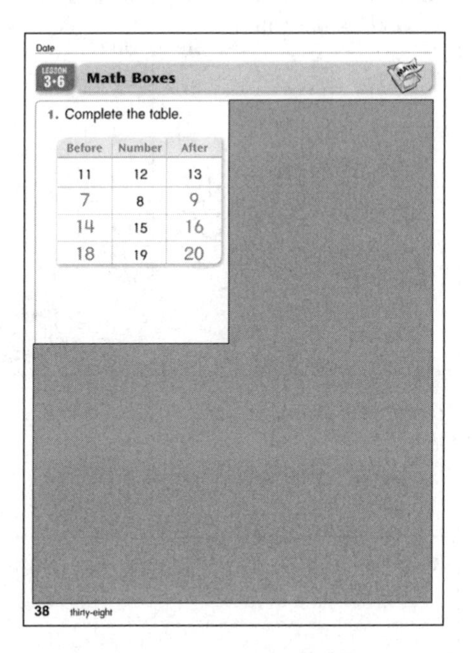

Students can focus on one Math Boxes problem at a time.

- ◆ Identify the problem or problems that are essential. Encourage students to complete these problems first. Suggest that students who finish a task early use their spare time to complete any unfinished Math Boxes problems. Consider providing time in your weekly schedule so that all students have the opportunity to complete unfinished Math Boxes problems.

- ◆ Have students complete the problems independently. Then have them form small groups and share their answers and explanations. As an alternative, ask students to complete the problems cooperatively even though the lesson indicates independent work.

- ◆ Divide the class into groups. Have each group solve one of the Math Boxes problems. "Jigsaw" to form new groups. Each of the new groups now has one student from each of the original groups. Each student in the new group is an expert on one of the problems. The expert explains the problem to the other students in the group.

- ◆ Have students complete Math Boxes pages as part of their daily morning routine. Math Boxes are one of the components of a lesson that lends itself to being completed outside of regular math time.

Modifying Math Boxes

The strategies suggested here can be used for many types of Math Boxes problems. The same types of modifications can be made to other journal pages as well.

- Modify the range of the numbers or ask students to record measurements to a more-precise or less-precise degree of accuracy to focus on a different level of a concept or skill.

2. Complete the "What's My Rule?" table and state the rule.

Rule: $\div 9$

in	out
45	5
81	9
27	3
36	4
72	8

SRB 162–166

To practice extended facts instead of basic facts, have students attach a zero to each number in the "What's My Rule?" table.

- Make various manipulatives, such as coins and bills, base-10 blocks, pattern blocks, and counters, available to provide concrete models for practicing concepts and skills.

6. Solve mentally or with a paper-and-pencil algorithm.

a. $3.56 + $2.49 = $6.05

b. $6.25 − $5.01 = $1.24

Encourage students to use coins and bills to help them solve decimal addition and subtraction problems.

- Have tools, such as number grids, place-value charts, calculators, or fact tables, available to help students solve problems.

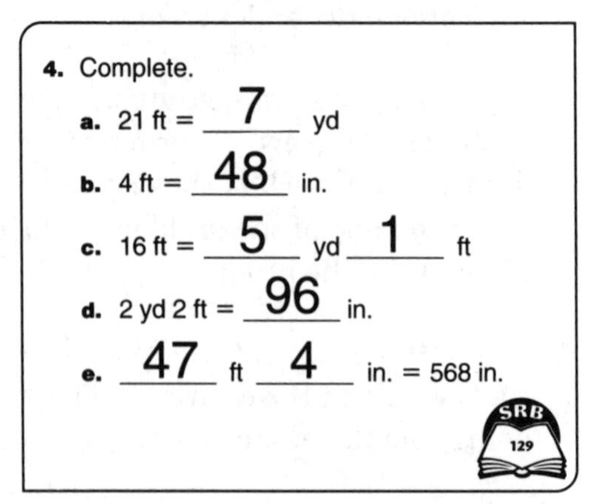

Some students may be able to solve Problems a–c mentally but choose to use a calculator with Problems d and e.

Creating Your Own Math Boxes

Occasionally, you may want to create your own Math Boxes page for practice or assessment purposes. There are blank masters on pages 136–141 of this handbook to serve this purpose. Consider the following ideas while designing pages for your class or individual students:

- Create a set of problems that focuses on a single concept or skill that students need to review, but address it in a variety of contexts. For example, focus on addition through number stories, facts problems, "What's My Rule?" problems, or skip counting. Because each Math Boxes page in the journal includes a variety of problems, each one targeting a different concept or skill, this strategy can help students who struggle with these transitions.

- Create a Math Boxes page that links with a set of Math Boxes pages in the journal. Tailor the numbers to meet the individual needs of students.

- Create a set of extra-practice problems in which all cells focus on concepts or skills from a particular lesson.

- Adapt *5-Minute Math* problems that address concepts or skills students need to review.

- Create a page in which each problem targets a specific concept or skill. Use the page for one week, each day replacing the numbers in the problems with new numbers.

- Use the templates of routines found on pages 138–141 of this handbook to create Math Boxes pages for students to complete. Fill in some of the numbers for each routine. Or, have students create the Math Boxes for classmates to complete. For more information about each of these routines, see the *Teacher's Reference Manual.*

- Have students write number stories in each cell of a template. Specify which operation should be the focus of each problem. Have students exchange Math Boxes pages and solve one another's problems.

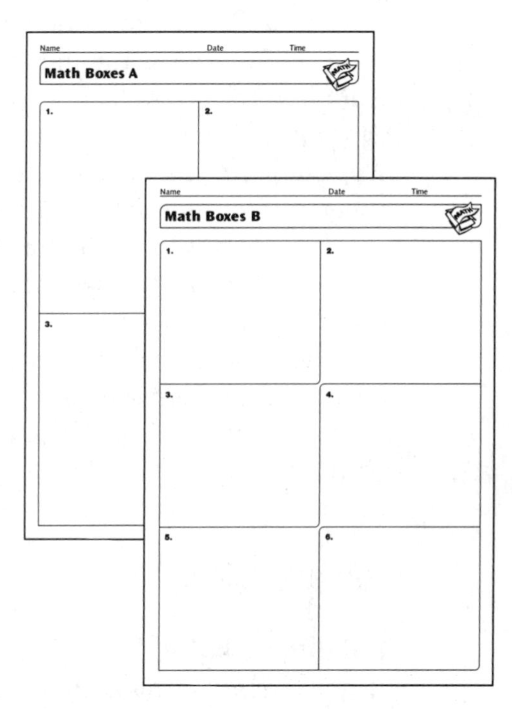

Differentiation Masters, pages 136 and 137, are templates for blank Math Boxes pages for four or six problems.

Differentiation Masters, page 138, is a master for a blank Math Boxes page that includes a variety of routines.

Using Part 3 of the Lesson

As written, *Everyday Mathematics* lessons engage a wide range of learners and support the development of mathematics concepts and skills at the highest possible level. There are times, however, when teachers still need to be flexible in implementing lessons. To address the individual needs of students, Part 3 of each lesson, Differentiation Options, provides additional resources beyond the scope of what is included in Part 1. The activities suggested are optional, intended to support rather than to replace lesson content. Many of the activities, designed so students can work with partners or small groups, are ideally suited for station work. Based on your professional judgment and assessments, determine when students might benefit from these activities. For each unit, use the master found on page 154 of this handbook to plan how you will use the Part 3 activities with the whole class, small groups, or individual students.

Part 3 Planning Master

Lesson	Readiness	Enrichment	Extra Practice	ELL Support
1-1		Whole class		Carlos Andres
1-2	Abby Conner Jamal	Chantel Eric	Matt Amy	
1-3			Cheryl DeAndre Melissa	Melanie
1-4	Toya Takako Kevin		Whole class	
1-5		Isabel Leon		Whole class
1-6	Whole class		Whole class	
1-7	Katherine Aman	Jayne	Hannah Abby Tom	Dmitry Carlos

Readiness Activities

Readiness activities introduce or develop the lesson content to support students as they work with the Key Concepts and Skills. Use Readiness activities with some or all students before teaching the lesson to preview the content so students are better prepared to engage in lesson activities. As an alternative, use Readiness activities at the completion of lesson activities to solidify students' understanding of lesson content.

In Lesson 3-1 of *Fourth Grade Everyday Mathematics,* students discuss problems in which one quantity depends on another. They illustrate this kind of relationship between pairs of numbers with a function machine and a "What's My Rule?" table. The following is the Readiness activity for the lesson.

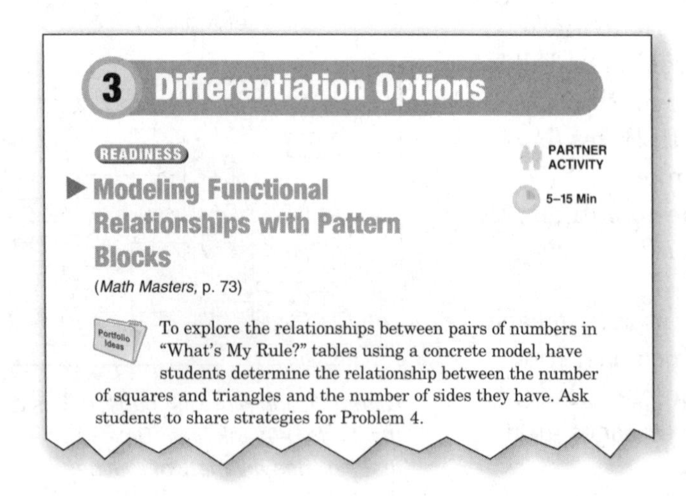

3 Differentiation Options

READINESS

PARTNER ACTIVITY

5–15 Min

▶ **Modeling Functional Relationships with Pattern Blocks**

(*Math Masters*, p. 73)

To explore the relationships between pairs of numbers in "What's My Rule?" tables using a concrete model, have students determine the relationship between the number of squares and triangles and the number of sides they have. Ask students to share strategies for Problem 4.

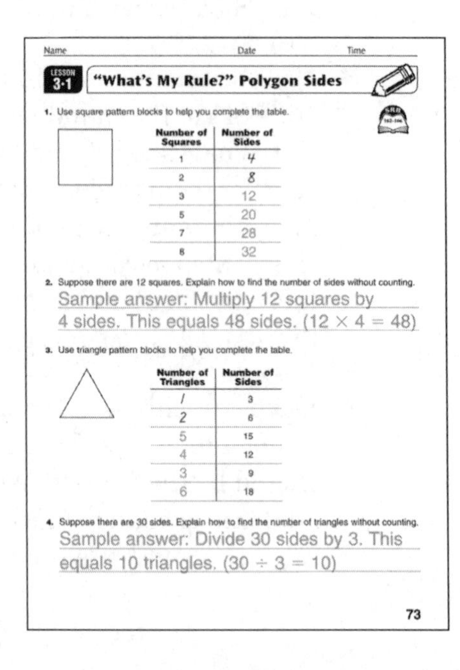

Name Date Time

LESSON 3-1 **"What's My Rule?" Polygon Sides**

1. Use square pattern blocks to help you complete the table.

Number of Squares	Number of Sides
1	4
2	8
3	12
5	20
7	28
8	32

2. Suppose there are 12 squares. Explain how to find the number of sides without counting.
Sample answer: Multiply 12 squares by 4 sides. This equals 48 sides. ($12 \times 4 = 48$)

3. Use triangle pattern blocks to help you complete the table.

Number of Triangles	Number of Sides
1	3
2	6
5	15
4	12
3	9
6	18

4. Suppose there are 30 sides. Explain how to find the number of triangles without counting.
Sample answer: Divide 30 sides by 3. This equals 10 triangles. ($30 \div 3 = 10$)

73

Enrichment Activities

Enrichment activities provide ways for students to apply or further explore Key Concepts and Skills emphasized in the lesson. Use the activities with some or all students after they have completed the lesson activities.

In Lesson 8-4 of *Second Grade Everyday Mathematics,* students explore the concept of equivalent fractions by matching fractional parts of circles. The following is the Enrichment activity for the lesson.

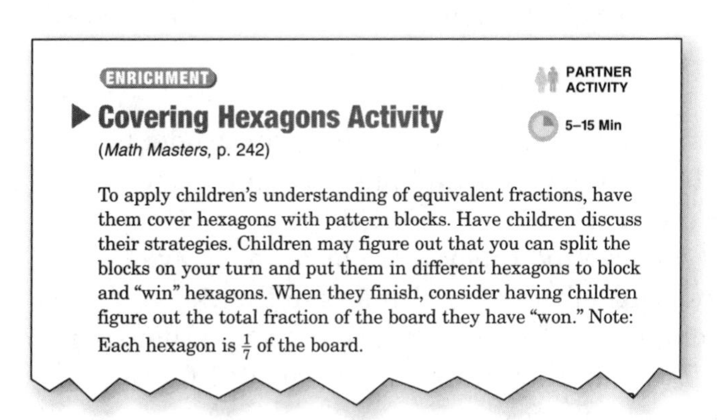

ENRICHMENT

PARTNER ACTIVITY

▶ **Covering Hexagons Activity**

5–15 Min

(*Math Masters,* p. 242)

To apply children's understanding of equivalent fractions, have them cover hexagons with pattern blocks. Have children discuss their strategies. Children may figure out that you can split the blocks on your turn and put them in different hexagons to block and "win" hexagons. When they finish, consider having children figure out the total fraction of the board they have "won." Note: Each hexagon is $\frac{1}{7}$ of the board.

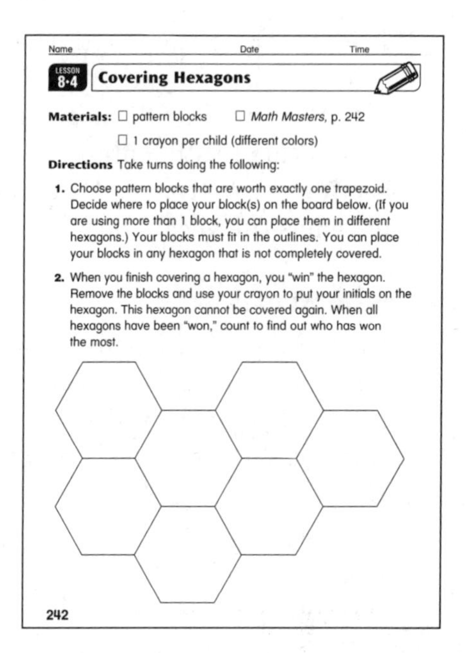

Name Date Time

LESSON 8·4 **Covering Hexagons**

Materials: ☐ pattern blocks ☐ *Math Masters,* p. 242

☐ 1 crayon per child (different colors)

Directions Take turns doing the following:

1. Choose pattern blocks that are worth exactly one trapezoid. Decide where to place your block(s) on the board below. (If you are using more than 1 block, you can place them in different hexagons.) Your blocks must fit in the outlines. You can place your blocks in any hexagon that is not completely covered.
2. When you finish covering a hexagon, you "win" the hexagon. Remove the blocks and use your crayon to put your initials on the hexagon. This hexagon cannot be covered again. When all hexagons have been "won," count to find out who has won the most.

242

Extra Practice Activities

These activities provide students with additional practice opportunities related to the content of the lesson. There are three main categories for extra practice activities—practice pages, games, and *5-Minute Math* problems.

The *5-Minute Math* problems are grouped by level of difficulty, which allows you to choose problems to meet the needs of your entire class or small groups of students. The activities can also serve as a catalyst for your own or students' problems and ideas.

5-Minute Math activities do the following:

- provide reinforcement and continuous review of Grade-Level Goals;
- provide practice with mental arithmetic and logical thinking activities;
- give students additional opportunities to think and talk about mathematics and to try out new ideas by themselves or with their teachers and classmates; and
- promote the process of solving problems, so in the long run, students become more willing to risk sharing their thoughts and their solution strategies with classmates rather than focus on getting quick answers.

Support for English Language Learners

The activities in the ELL Support section are designed to promote development of language related to Key Concepts and Skills. Several vocabulary routines are established early in each grade and are revisited throughout the year.

It is important to note that English language learners should not be restricted solely to ELL Support activities. Often the Readiness and Enrichment activities are ideally suited to enhance the mathematical content for this population of students. Likewise, although the activities described in this section are extremely helpful for English language learners, this kind of work enriches the vocabulary development of all students.

Math Word Bank

Use a Math Word Bank, which is similar to a dictionary, to invite students to make connections between new terms and words and phrases they know. For each entry, have students make a visual representation of the word or phrase and list three related terms that will remind them of the meaning. Have English language learners record some of the related words in their own language. Have students keep completed pages in a 3-ring binder so that they may refer to them as necessary. Two different masters are provided for this routine on pages 142 and 143 of this handbook.

In Lesson 2-1 of *Sixth Grade Everyday Mathematics,* students explore reading and writing large numbers. The following is the ELL Support activity for this lesson.

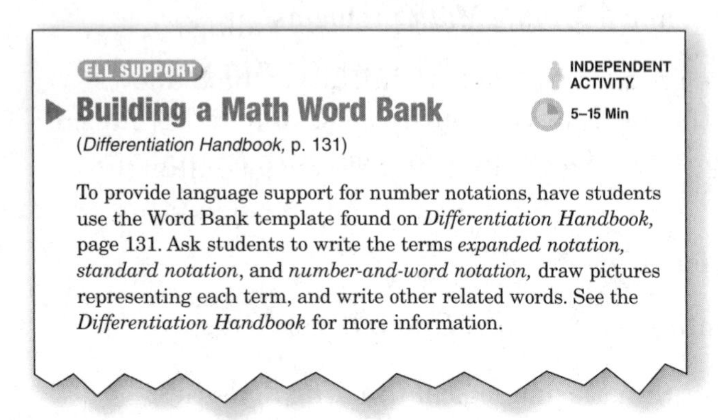
ELL SUPPORT

INDEPENDENT ACTIVITY

5–15 Min

▶ Building a Math Word Bank

(*Differentiation Handbook,* p. 131)

To provide language support for number notations, have students use the Word Bank template found on *Differentiation Handbook,* page 131. Ask students to write the terms *expanded notation, standard notation,* and *number-and-word notation,* draw pictures representing each term, and write other related words. See the *Differentiation Handbook* for more information.

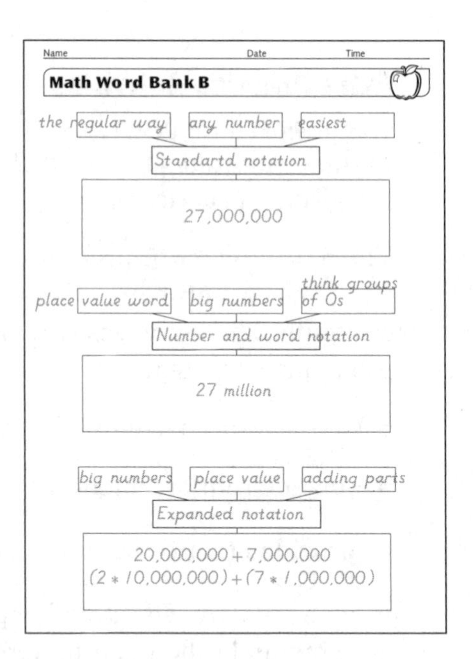
Name Date Time

Math Word Bank B

the regular way | any number | easiest

Standartd notation

27,000,000

place value word | big numbers | think groups of 0s

Number and word notation

27 million

big numbers | place value | adding parts

Expanded notation

20,000,000 + 7,000,000
(2 * 10,000,000) + (7 * 1,000,000)

Museums

In *Everyday Mathematics,* museums help students connect the mathematics they are studying with their everyday lives. A museum is simply a collection of objects, pictures, or numbers that illustrates or incorporates mathematical concepts related to the lessons. Museums provide opportunities for students to explore and discuss new mathematical ideas. If several English language learners speak the same language, have them take a minute to discuss museums in their own language first and then share in English as they are able.

In Lesson 5-7 of *Second Grade Everyday Mathematics,* students construct pyramids using straws and connectors. They discuss the properties of the pyramids they have built. The following is the ELL Support activity for this lesson.

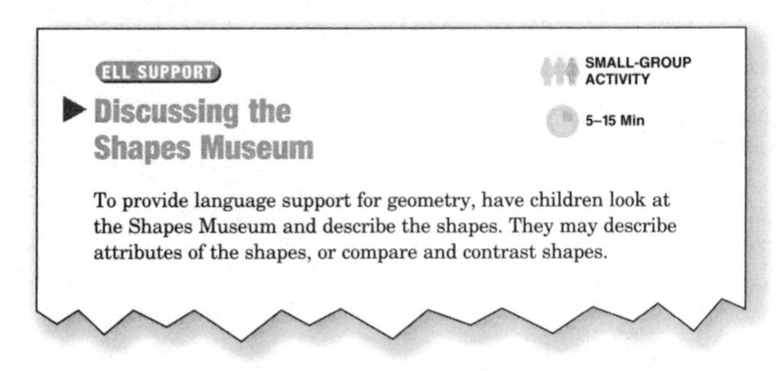
ELL SUPPORT

SMALL-GROUP ACTIVITY

5–15 Min

▶ Discussing the Shapes Museum

To provide language support for geometry, have children look at the Shapes Museum and describe the shapes. They may describe attributes of the shapes, or compare and contrast shapes.

Student- and Teacher-Made Posters

Sometimes a unit focuses on a topic that introduces potentially confusing content, for example, a great deal of new vocabulary, many steps in a problem-solving process, or several strategies for solving a problem. Providing students with a poster to use as a reference or having them create their own posters can help them make sense of such complex content.

In Unit 4 of *Fifth Grade Everyday Mathematics,* students review long-division algorithms. The following is the ELL Support activity for Lesson 4-5 in this unit.

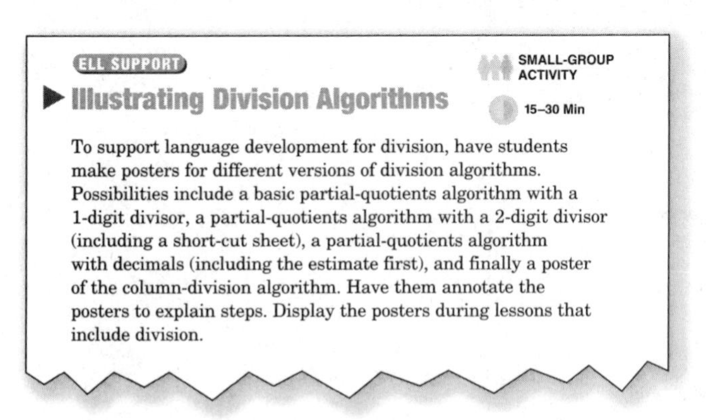
ELL SUPPORT

SMALL-GROUP ACTIVITY

15–30 Min

▶ Illustrating Division Algorithms

To support language development for division, have students make posters for different versions of division algorithms. Possibilities include a basic partial-quotients algorithm with a 1-digit divisor, a partial-quotients algorithm with a 2-digit divisor (including a short-cut sheet), a partial-quotients algorithm with decimals (including the estimate first), and finally a poster of the column-division algorithm. Have them annotate the posters to explain steps. Display the posters during lessons that include division.

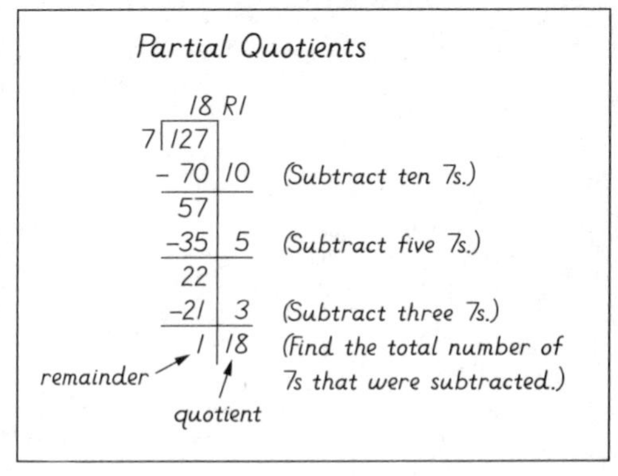

Everyday Mathematics also includes posters, such as the Probability Meter Poster. When you use the *Everyday Mathematics* posters with English language learners, you may display both the English and Spanish versions simultaneously or only the English version.

Graphic Organizers

Students find it easier to learn and retain new words if they connect the new words to their existing vocabulary. Graphic organizers are organizational tools for making connections more explicit and for helping students gain a deeper understanding of a concept.

In Lesson 6-10 of *Second Grade Everyday Mathematics,* students explore multiplication and division and the relationships between these operations. The following is the ELL Support activity for this lesson.

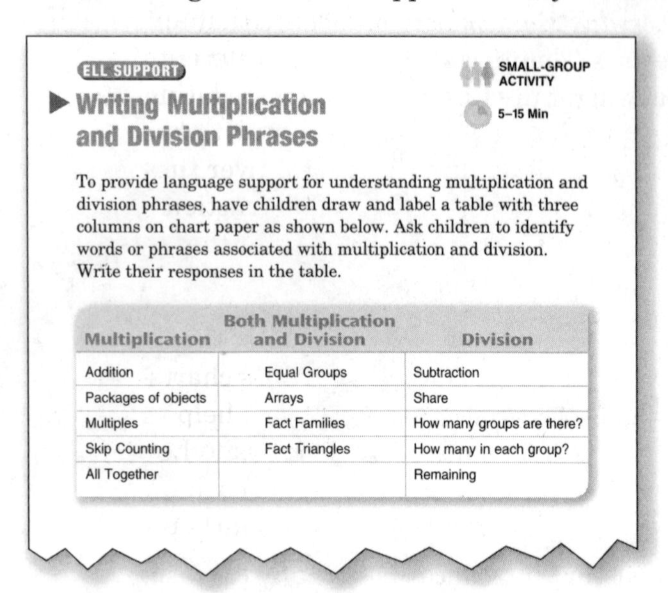

ELL SUPPORT

SMALL-GROUP ACTIVITY

5–15 Min

▶ Writing Multiplication and Division Phrases

To provide language support for understanding multiplication and division phrases, have children draw and label a table with three columns on chart paper as shown below. Ask children to identify words or phrases associated with multiplication and division. Write their responses in the table.

Multiplication	Both Multiplication and Division	Division
Addition	Equal Groups	Subtraction
Packages of objects	Arrays	Share
Multiples	Fact Families	How many groups are there?
Skip Counting	Fact Triangles	How many in each group?
All Together		Remaining

In Lesson 8-7 of *Sixth Grade Everyday Mathematics,* students use ratios as a strategy for solving percent problems. The following is the ELL Support activity for this lesson.

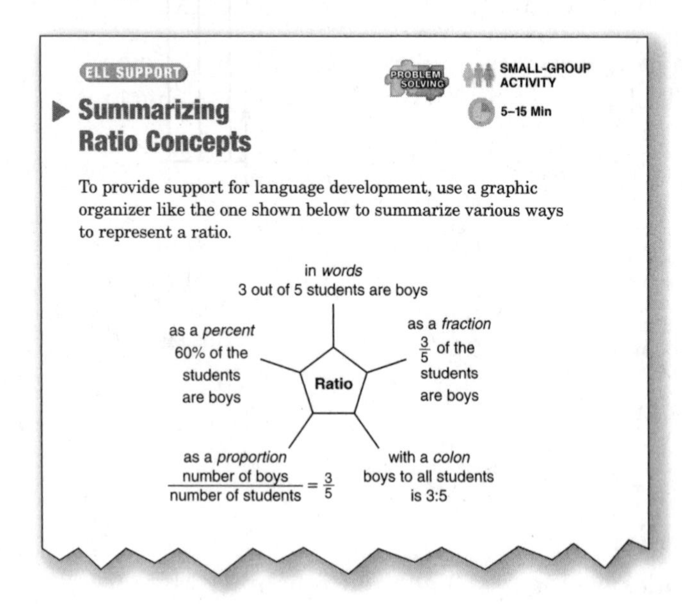

ELL SUPPORT

PROBLEM SOLVING

SMALL-GROUP ACTIVITY

5–15 Min

▶ Summarizing Ratio Concepts

To provide support for language development, use a graphic organizer like the one shown below to summarize various ways to represent a ratio.

Looking at Grade-Level Goals

Students using *Everyday Mathematics* are expected to master a great deal of mathematical content, but not necessarily the first time the concepts and skills are introduced. The *Everyday Mathematics* curriculum aims for proficiency through repeated exposure over several years of study.

All of the content in *Everyday Mathematics* is important, whether it's being treated for the first time or the fifth time. The *Everyday Mathematics* curriculum is like an intricately woven rug with many threads that appear and reappear to form complex patterns. Students will progress at different rates, so multiple exposures to important content are critical for accommodating individual differences. The program is created to be consistent with how students actually learn mathematics, building understanding over time, first through informal exposure and later through more-formal instruction. It is crucial that students have the opportunity to experience all that the curriculum has to offer in every grade.

To understand where concepts and skills are revisited over time, the unit-specific section of this handbook, beginning on page 49, includes charts for looking at the Grade-Level Goals in each unit. These charts will help you see where you are in the development of the goals—whether the Grade-Level Goal is taught, practiced and applied, or not a focus in the lessons of each unit. The excerpt below can help you understand the information these charts provide.

Map of Number and Numeration Goal 1 for Fourth Grade

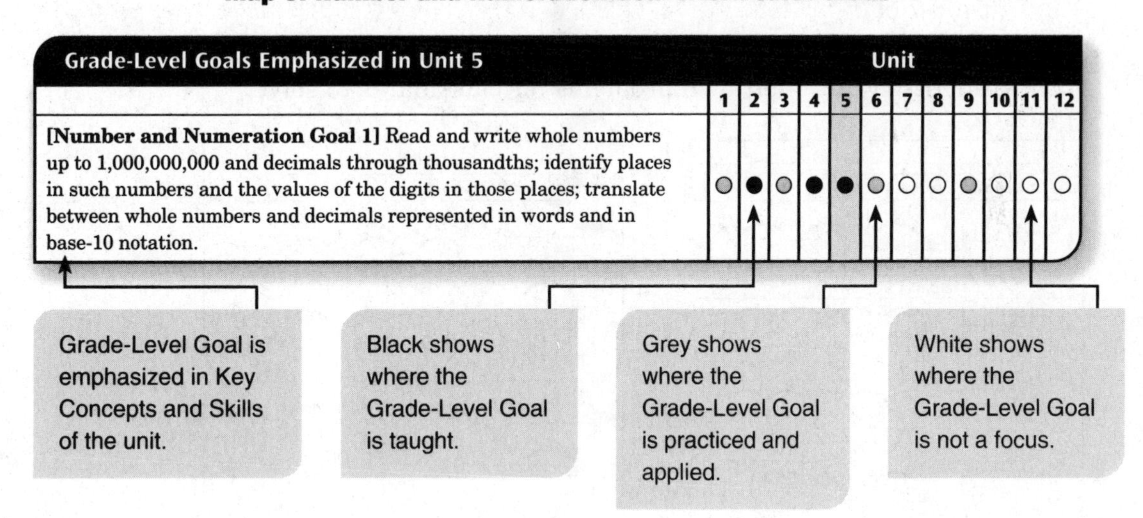

Maintaining Concepts and Skills

The charts discussed in the previous section illustrate where each Grade-Level Goal in a unit is revisited over the course of the year. Sometimes there will be several units in a row that do not address a Grade-Level Goal through either the Key Concepts and Skills emphasized in lessons or through practice. Moreover, as the year progresses, some goals reach the end of their formal development at that grade level. Because students progress at different rates, you may sometimes have students who need to revisit concepts and skills for a particular Grade-Level Goal.

At the end of each unit overview in the unit-specific section of this handbook, you will find a list of *Maintaining Concepts and Skills* activities that you can use to provide students with additional opportunities to explore, review, or practice content. Frequently the list will include references to program routines. These routines, which are revisited throughout the curriculum across the grades, provide a comfortable and convenient way to reinforce, maintain, or further develop concepts and skills for individual students. Blank masters for these routines are included in this handbook beginning on page 135. Examples of helpful strategies are described here.

Frames and Arrows

This routine, which is emphasized in Grades 1 through 3, provides opportunities for students to practice basic and extended addition, subtraction, and multiplication facts. The problems also require students to use algebraic thinking involving patterns, functions, and sequences.

Use these masters to create pages to meet the needs of individual students, or have students create their own problems for classmates to solve.

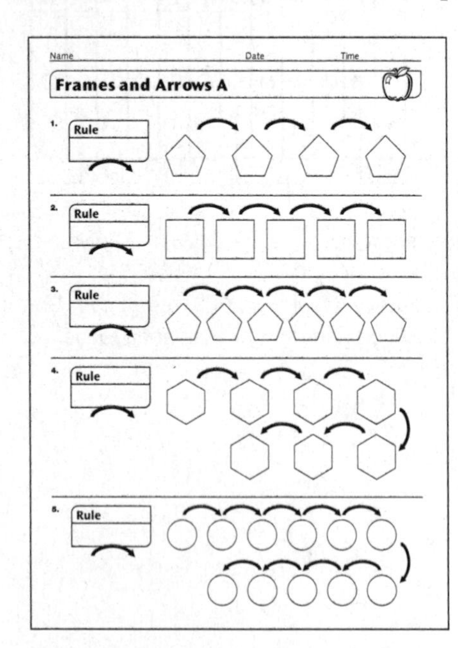

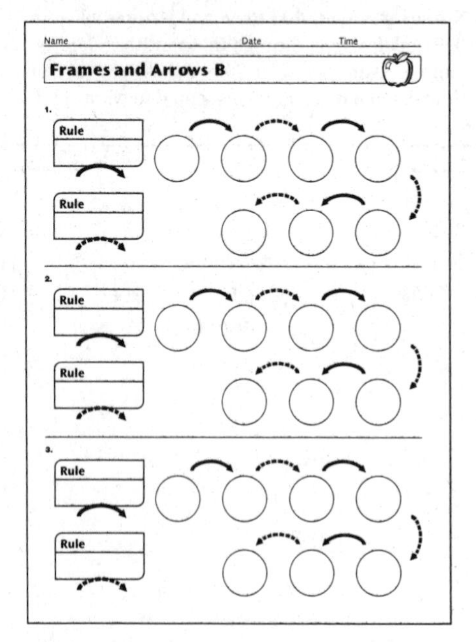

Students use page 144, to practice solving Frames-and-Arrows problems with one rule, and page 145, to practice solving Frames-and-Arrows problems with two rules.

"What's My Rule?"

This routine, which is introduced in Grade 1 and continues through Grade 6, provides opportunities for students to practice basic computation skills and solve problems involving functions. In the upper grades, the functions can be represented visually in graphs and algebraically using variables. In addition to solving teacher-generated problems, students can generate problems for one another to solve. There is a blank master for "What's My Rule?" on page 146 of this handbook.

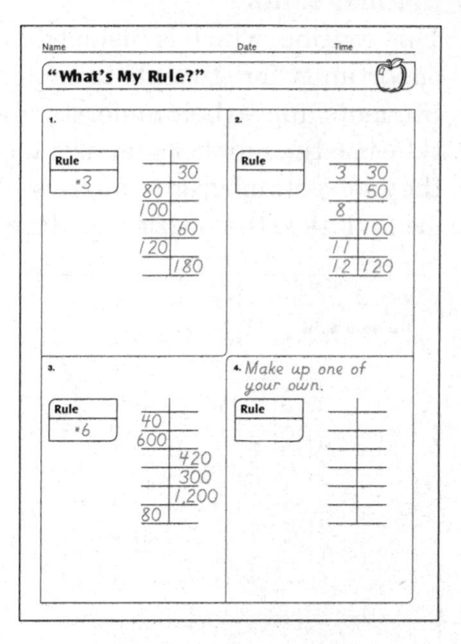
Name Date Time

"What's My Rule?"

1. Rule: *3

in	out
	30
80	
100	
	60
120	
	180

2. Rule:

in	out
3	30
	50
8	
	100
11	
12	120

3. Rule: *6

in	out
40	
600	
	420
	300
	1,200
80	

4. Make up one of your own.
Rule:

A teacher-generated set of problems that focuses on basic and extended multiplication facts

Name-Collection Boxes

This routine, which is used in Grades 1 through 6, provides the opportunity for students to practice basic computation skills, generate equivalent names for numbers, use grouping symbols, and apply order of operations to numerical expressions. In addition to solving teacher-generated problems, students can generate problems for one another to solve. There is a blank master for name-collection boxes on page 147 of this handbook.

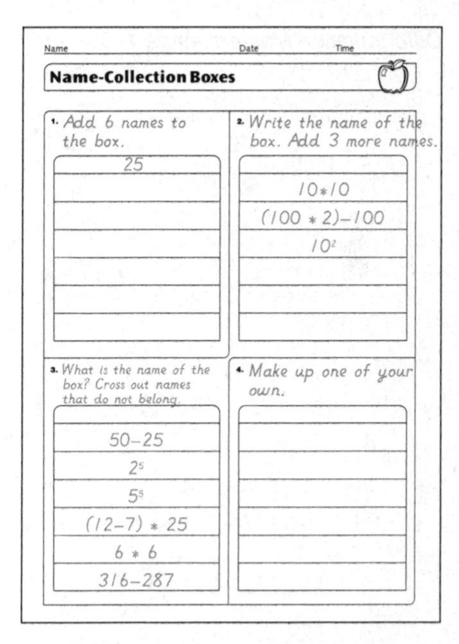
Name Date Time

Name-Collection Boxes

1. Add 6 names to the box.
25

2. Write the name of the box. Add 3 more names.
10*10
(100 * 2)–100
10²

3. What is the name of the box? Cross out names that do not belong.
50–25
2⁵
5⁵
(12–7) * 25
6 * 6
316–287

4. Make up one of your own.

A teacher-generated page that illustrates the variety of ways in which name-collection box problems can be formatted

Number Grids

This routine, which is included in Kindergarten through Grade 3, provides the opportunity for students to explore number relationships and number patterns. Students apply their understanding of these patterns and relationships when they use the number grid as a tool for solving computation problems and when they solve number-grid puzzles. When students become familiar with number-grid puzzles, the puzzles can be extended to include any range of numbers.

Name Date Time

Number Grid

-9	-8	-7	-6	-5	-4	-3	-2	-1	0
1	2	3	4	5	6	7	8	9	10
11	12	13	14	15	16	17	18	19	20
21	22	23	24	25	26	27	28	29	30
31	32	33	34	35	36	37	38	39	40
41	42	43	44	45	46	47	48	49	50
51	52	53	54	55	56	57	58	59	60
61	62	63	64	65	66	67	68	69	70
71	72	73	74	75	76	77	78	79	80
81	82	83	84	85	86	87	88	89	90
91	92	93	94	95	96	97	98	99	100
101	102	103	104	105	106	107	108	109	110

-9	-8	-7	-6	-5	-4	-3	-2	-1	0
1	2	3	4	5	6	7	8	9	10
11	12	13	14	15	16	17	18	19	20
21	22	23	24	25	26	27	28	29	30
31	32	33	34	35	36	37	38	39	40
41	42	43	44	45	46	47	48	49	50
51	52	53	54	55	56	57	58	59	60
61	62	63	64	65	66	67	68	69	70
71	72	73	74	75	76	77	78	79	80
81	82	83	84	85	86	87	88	89	90
91	92	93	94	95	96	97	98	99	100
101	102	103	104	105	106	107	108	109	110

Differentiation Masters, page 148

66	67	
	77	78
	87	
96	97	
	107	108

A teacher-generated number-grid puzzle that starts with 66 and 87

991	992	
	1,002	1,003
1,011		1,013
1,021		1,023
1,031		**1,033**

A teacher-generated number-grid puzzle that starts with 1,002 and 1,033

Projects

This section offers suggestions for how to differentiate the Grade 5 Projects for your students. For each project, you will find three differentiation options: Adjusting the Activity Ideas, ELL Support, and a suggested Writing/Reasoning prompt.

Contents

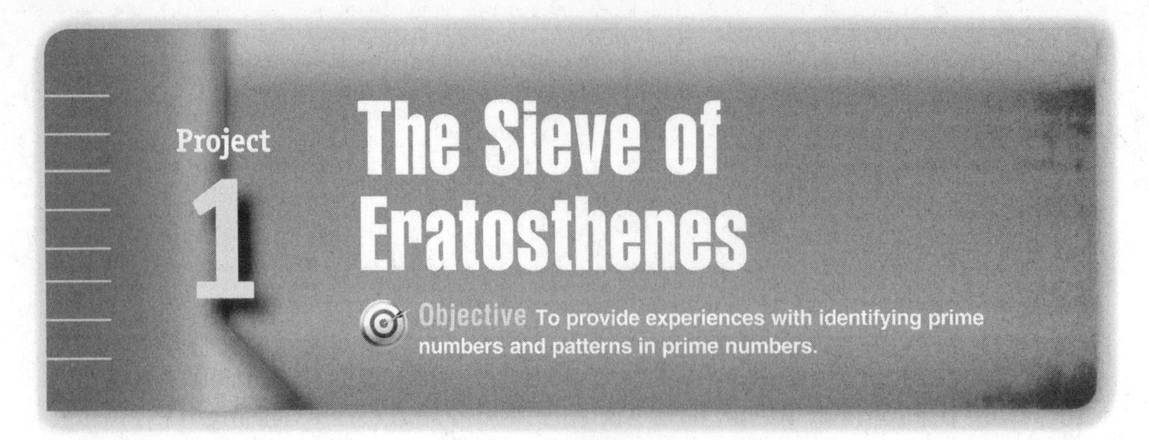

Projects are located at the back of both Volume 1 and Volume 2 of the Grade 5 *Teacher's Lesson Guide.* Use this project during or after Unit 1.

Adjusting the Activity Ideas

- Have students use Multiplication/Division Facts Tables or calculator skip counting to identify multiples.
- Have students complete this same activity for the numbers 1,000 to 1,100.
- Have students find the prime numbers from 3 to 53 that can be formed by the sum of two perfect squares. A perfect square is a number that can be renamed as n^2 where n is a counting number.

ELL Support

To provide language support for the Sieve of Eratosthenes project, have students use the Word Bank Template found on page 142 in this handbook. Ask students to write the terms *prime* and *composite,* draw pictures relating to each term, and write other related words. See page 32 of this handbook for more information.

Writing/Reasoning

Have students write a response to the following: Explain how you would determine whether the number 151 is prime or composite. *(Sample answer: I have to try to find all of its factors. If 1 and 151 are the only two factors, then I know it is a prime number.)*

Project 2

Deficient, Abundant, and Perfect Numbers

Objective To explore number properties.

Projects are located at the back of both Volume 1 and Volume 2 of the Grade 5 *Teacher's Lesson Guide.* Use this project during or after Unit 1.

Adjusting the Activity Ideas

- Have students research how to identify "happy numbers" (counting numbers for which the sum of the squares of the digits eventually ends in 1), and find the 20 happy numbers in the first 100 counting numbers. For example, 13 is happy because $1^2 + 3^2 = 10$ and $1^2 + 0^2 = 1$.
- Have students use Multiplication/Division Facts Tables or calculator skip counting to identify multiples.
- Have students make factor rainbows for the numbers.

ELL Support

To provide language support for the Deficient, Abundant, and Perfect Numbers project, have students use the Word Bank Template found on page 143 in this handbook. Ask students to write the terms *deficient numbers, abundant numbers,* and *perfect numbers;* draw pictures relating to each term; and write other related words. See page 32 of this handbook for more information.

Writing/Reasoning

Have students write a response to the following: How does organizing whole numbers by the sums of their proper factors help you to play *Factor Captor? (Sample answer: I can use my list to select the best possible numbers so I can have the highest possible score. I can pick large numbers that have very low sums of proper factors so my partner gets fewer points than I do.)*

Projects are located at the back of both Volume 1 and Volume 2 of the Grade 5 *Teacher's Lesson Guide.* Use this project during or after Unit 2.

Adjusting the Activity Ideas

- After students use the Egyptian Algorithm to solve the problems on *Math Masters,* page 384, have them write an explanation for how the algorithm works.
- Have students make a chart of Roman numerals for 0–100.
- Have students first solve the problems using a method of their choice before trying to use the Egyptian algorithm.

ELL Support

To provide language support for the Egyptian Algorithm project, have students make posters for all of the multiplication algorithms they know. Each student or group of students can make one poster for a different algorithm. They can make posters for mental-math algorithms as well as paper-and-pencil algorithms. For example, the mental math strategy of 99 * 28 is 28 less than 100 * 28 could be illustrated step by step on a poster. Display the posters and have students discuss advantages and disadvantages for each strategy.

Writing/Reasoning

Have students write a response to the following: How is the other ancient algorithm similar to and different from the Egyptian algorithm? *(Sample answer: They are alike because you double one of the factors. They are different because in the Egyptian algorithm, you end up with a total that equals one of the factors and in the other algorithm you do not.)*

Projects are located at the back of both Volume 1 and Volume 2 of the Grade 5 *Teacher's Lesson Guide.* Use this project during or after Unit 2.

Adjusting the Activity Ideas

- For Super Speedy Addition, have students complete the computation trick using 2-digit numbers instead. The total they make is 99.
- Have students explain which trick they think is the most difficult to figure out and why.
- Have students figure out how to perform the Crazy Calendar Addition trick on a number grid instead of a calendar. Have them explain how it would be different and how it would be the same on the number grid.

ELL Support

To provide language support for the Magic Trick project, have students use the Word Bank Template found on page 142 in this handbook. Ask students to write the terms *row* and *column,* draw pictures relating to each term, and write other related words. See page 32 of this handbook for more information.

Writing/Reasoning

Have students write a response to the following: How can the Super Speedy Addition trick work if your partner were to give you five 3-digit numbers? *(Sample answer: If my partner gave me five different 3-digit numbers, I would add a number to my partner's first four 3-digit numbers to make each sum equal 999. I know that the sum of these eight numbers will equal 4 less than 4,000, so I would subtract 4 from my partner's fifth number and then add that result to 4,000.)*

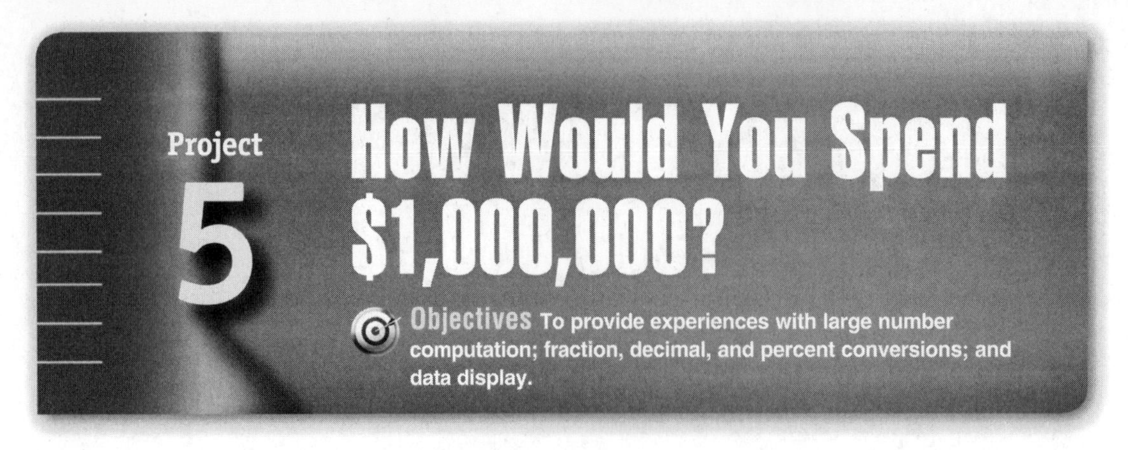

Projects are located at the back of both Volume 1 and Volume 2 of the Grade 5 *Teacher's Lesson Guide.* Use this project during or after Unit 5.

Adjusting the Activity Ideas

- Have students spend $1,000 or $10,000.
- Have students convert the million dollars into another currency and change the totals of their completed project accordingly. Another option is to have them choose another currency and use a million of that currency to see the difference in buying power compared to the dollar.
- Have students look at the budget of the government or a business and create a circle graph showing how much money is spent and in what areas.

ELL Support

- To provide language support for the Million Dollar project, have students use the Word Bank Template found on page 142 in this handbook. Ask students to write the term *unit price,* draw pictures relating to the term, and write other related words. See page 32 of this handbook for more information.
- Have students compare the terms *unit fraction* and *unit price* and describe what the terms have in common.

Writing/Reasoning

Have students write a response to the following: Was it more difficult to spend $1,000,000 than you imagined? *(Sample answer: I thought it was going to be very easy but it turned out to be a lot harder. After researching and recording how the money was spent, I started to realize all the different items that I would need for my project. I realized it would take more time and I might have money left over.)*

Project

6 Sports Areas

Objectives To provide experiences with calculating the areas of rectangles, converting between units, and converting mixed units to decimals.

Projects are located at the back of both Volume 1 and Volume 2 of the Grade 5 *Teacher's Lesson Guide.* Use this project during or after Unit 9.

Adjusting the Activity Ideas

- Have small groups of students measure the school play area or field to figure the square footage. Then have them determine if the play area for the school is more or less than one acre.
- Have students make a bar graph to compare the square footage of the playing areas for the five contact sports.
- For each area given on *Math Masters,* pages 401 and 402, have students find out how many players are involved on each play area and then calculate the play area square footage per person and comment on possible reasons for this ratio.

ELL Support

- To provide language support for the Sports Areas project, have students use the Word Bank Template found on page 142 in this handbook. Ask students to write the term *scale drawing,* draw pictures relating to the term, and write other related words. See page 32 of this handbook for more information.
- Have students discuss the various meanings of the word *scale* and its meaning in the context of this project.

Writing/Reasoning

Have students write a response to the following: How many tennis courts can fit on a football field? Hint: There will be space left over because the shape of the court must remain the same. *(Sample answer: 16. I first made a drawing of the football field and then sketched tennis courts inside it. I sketched the tennis courts so the longer side was horizontal and I could fit 8. Then I turned them so the longer side was vertical. I was able to fit 16. If you try to divide the area of the football field by the area of the tennis courts, you get 17—but that doesn't account for the shape of the courts.)*

Projects are located at the back of both Volume 1 and Volume 2 of the Grade 5 *Teacher's Lesson Guide.* Use this project during or after Unit 9.

Adjusting the Activity Ideas

- Have students use Pick's Formula to calculate the area of more familiar polygons such as squares, rectangles, and triangles.
- Have partners create various polygons on geoboards and calculate the area using Pick's Formula.
- Have students create a page for a Pick's Formula scrapbook. Each student creates an appropriate polygon and puts the area on the bottom with a stick-on note covering the answer. Use 1-cm grid paper cut in half to limit the size of the polygons.

ELL Support

To provide language support for the Pick's Formula project, have students use the Word Bank Template found on page 142 in this handbook. Ask students to write the terms *vertices* and *interior,* draw pictures relating to each term, and write other related words. See page 32 of this handbook for more information.

Writing/Reasoning

Have students write a response to the following: What are some of the advantages and disadvantages to Pick's Formula? *(Sample answer: If the polygon is drawn on grid paper, it is easy to find the area of any polygon. If the polygon is not drawn on grid paper, there is no way to use Pick's Formula.)*

Project

8 Pendulums

Objective To provide experiences with collecting, displaying, and analyzing data.

Projects are located at the back of both Volume 1 and Volume 2 of the Grade 5 *Teacher's Lesson Guide.* Use this project during or after Unit 10.

Adjusting the Activity Ideas

- Have students research Newton's Cradle and make a Venn diagram comparing and contrasting this to a regular pendulum.
- Have students explain why they are timing ten complete swings.
- Have students make a virtual pendulum swing and read why the pendulum reacts as it does at http://pbskids.org/zoom/games/pendulum/.

ELL Support

To provide language support for the Pendulum project, have students use the Word Bank Template found on page 142 in this handbook. Ask students to write the term *pendulum,* draw pictures relating to the term, and write other related words. See page 32 of this handbook for more information.

Writing/Reasoning

Have students write a response to the following: According to the chart, how long should the pendulum in a clock be for the pendulum to complete one swing every second? *(Sample answer: At 20 cm, it completes 1 swing in 0.9 sec and at 30 cm it completes 1 swing in 1.1 seconds. Therefore, I would make the string 25 cm long because 1.0 second is halfway between 0.9 and 1.1, and 25 cm is halfway between 20 and 30.)*

Project 9

Adding Volumes of Solid Figures

Objective To find the volume of solid figures composed of two non-overlapping rectangular prisms.

Projects are located at the back of both Volume 1 and Volume 2 of the Grade 5 *Teacher's Lesson Guide.* Use this project during or after Unit 9.

Adjusting the Activity Ideas

- Provide students with additional open boxes to use for practice estimating volume when the boxes are filled with centimeter cubes.
- Ask students to work in groups of three to find the volume of three non-overlapping rectangular prisms.
- Ask students to list real-life examples of adjoining, non-overlapping rectangular prisms.

ELL Support

- To provide language support for the Adding Volumes of Solid Figures project, have students use the Word Bank Template found on page 142 of this handbook. Ask students to write the term *unit cube,* draw pictures relating to the term, and write other related words. See page 32 of this handbook for more information.
- Record the two formulas that may be used to find the volume of a rectangular prism on the class data pad: $V = B * h$ or $V = l * w * h$. Ask students to explain each of the formulas in their own words.
- Have students discuss the various everyday meanings of the word *volume* and its meaning in the context of this project.

Writing/Reasoning

Have students write a response to the following: How could you find the approximate volume of your classroom? (*Sample answer: I would first measure the approximate length and width of the classroom. I would then multiply the two measures to get the approximate area of the classroom floor. I would then measure the height of the classroom from floor to ceiling. This measure would give me the height. I would then multiply the area of the room* (B) *by the height* (h), *or* B * h, *to get the approximate volume of the classroom.*)

Algorithm Project 1, U.S. Traditional Addition, along with all other Grade 5 projects, is located at the back of both Volume 1 and Volume 2 of the Grade 5 *Teacher's Lesson Guide.* Use this project after Lesson 2-2.

Adjusting the Activity Ideas

- Have students start with a 2-digit example to allow them to focus on the procedure.
- Have students use a place-value chart when using this method to keep track of the values of the digits.
- Have students use base-10 blocks to illustrate each step of U.S. traditional addition. Use base-10 blocks either before or after recording the steps on paper.

ELL Support

Have students mark the appropriate columns with the place-value terms ones, tens, hundreds, and thousands. This will help them to remember the values of the digits they are adding.

Writing/Reasoning

Have students write a response to the following: How would you explain this way of adding to someone else? *(Sample answer: I would tell them that you need to work from right to left. Imagine you are adding two 2-digit numbers. First add the digits in the ones column. If the answer is only one digit, you write it underneath the line. If the answer is two digits, you write the ones digit underneath the line and the tens digit above the tens column because this is really a ten. Then add the digits in the tens column. If there is a digit above, you need to add that in. Write the ones digit of this sum in the tens column underneath because this is really a ten, and if there is a tens digit write it in the hundreds column underneath because this is really a hundred. The number below the line is the answer.)*

Algorithm 2 Project

U.S. Traditional Addition: Decimals

Objective To introduce U.S. traditional addition for decimals.

Algorithm Project 2, U.S. Traditional Addition: Decimals, along with all other Grade 5 projects, is located at the back of both Volume 1 and Volume 2 of the Grade 5 *Teacher's Lesson Guide.* Use this project after Lesson 2-2.

Adjusting the Activity Ideas

- Have students start with a 2-digit example to allow them to focus on the procedure.
- Have students use a place-value chart when using this method to keep track of the values of the digits and line up the decimal points.
- Have students use base-10 blocks to illustrate each step of U.S. traditional addition. Use base-10 blocks either before or after recording the steps on paper.

ELL Support

Have students mark the appropriate columns with the place-value terms ones, tens, hundreds, and, to the right of the decimal point, tenths, hundredths, and thousandths. This will help them to remember the values of the digits they are adding.

Writing/Reasoning

Have students write a response to the following: How is adding with decimals different from adding with whole numbers? *(Sample answer: Adding with decimals is different from adding with whole numbers because you need to line up the decimal points before you start adding. After you finish adding you need to put the decimal point in the answer by lining it up with the other decimal points.)*

Algorithm
3
Project

U.S. Traditional Subtraction

Objective To introduce U.S. traditional subtraction.

Algorithm Project 3, U.S. Traditional Subtraction, along with all other Grade 5 projects, is located at the back of both Volume 1 and Volume 2 of the Grade 5 *Teacher's Lesson Guide*. Use this project after Lesson 2-3.

Adjusting the Activity Ideas

- Have students start with a 2-digit example to allow them to focus on the procedure.
- Have students use a place-value chart when using this method to keep track of the values of the digits.
- Have students use base-10 blocks to illustrate each step of U.S. traditional subtraction. Use base-10 blocks either before or after recording the steps on paper.

ELL Support

Have students mark the appropriate columns with the place-value terms ones, tens, and hundreds. This will help them to remember the values of the digits they are subtracting.

Writing/Reasoning

Have students write a response to the following: How would you explain this way of subtracting to someone else? *(Sample answer: I would tell them that you need to work from right to left. Instead of doing all the trades beforehand, you check if you need to regroup as you go and subtract the numbers in one column at a time.)*

Algorithm 4 Project

U.S. Traditional Subtraction: Decimals

Objective To introduce U.S. traditional subtraction with decimals.

Algorithm Project 4, U.S. Traditional Subtraction: Decimals, along with all other Grade 5 projects, is located at the back of both Volume 1 and Volume 2 of the Grade 5 *Teacher's Lesson Guide*. Use this project after Lesson 2-3.

Adjusting the Activity Ideas

- Have students start with a 2-digit example to allow them to focus on the procedure.
- Have students use a place-value chart when using this method to keep track of the values of the digits and line up the decimal points.
- Have students use base-10 blocks to illustrate each step of U.S. traditional subtraction. Use base-10 blocks either before or after recording the steps on paper.

ELL Support

Have students mark the appropriate columns with the place-value terms ones, tens, hundreds, and, to the right of the decimal point, tenths and hundredths. This will help them to remember the values of the digits they are subtracting.

Writing/Reasoning

Have students write a response to the following: How would you explain this way of subtracting to someone else? *(Sample answer: I would tell them that you need to work from right to left and make sure the decimal points are lined up. Instead of doing all the trades beforehand, you check if you need to regroup as you go and subtract the numbers in one column at a time. After you finish subtracting, you need to put the decimal point in the answer by lining it up with the other decimal points.)*

Algorithm Project 5, U.S. Traditional Multiplication, along with all other Grade 5 projects, is located at the back of both Volume 1 and Volume 2 of the Grade 5 *Teacher's Lesson Guide.* Use this project after Lesson 2-9.

Adjusting the Activity Ideas

- Have students use a place-value chart when using this method to keep track of the values of the digits.
- Have students use a multiplication/division facts table and their knowledge of fact extensions to help them solve problems using U.S. traditional multiplication.
- Have students model each step of U.S. traditional multiplication by using the partial-products method simultaneously.

ELL Support

Have students mark the appropriate columns with the place-value terms ones, tens, and hundreds. This will help them to remember the values of the digits they are multiplying.

Writing/Reasoning

Have students write a response to the following: How would you explain this way of multiplying to someone else? *(Sample answer: It is similar to partial products except you multiply from right to left. Instead of writing each product on a separate line, you write each digit of the partial products in the appropriate place-value column.)*

Algorithm 6 Project

U.S. Traditional Multiplication: Decimals

Objective To introduce U.S. traditional multiplication using decimals.

Algorithm Project 6, U.S. Traditional Multiplication: Decimals, along with all other Grade 5 projects, is located at the back of both Volume 1 and Volume 2 of the Grade 5 *Teacher's Lesson Guide*. Use this project after Lesson 2-9.

Adjusting the Activity Ideas

- Have students use a place-value chart when using this method to keep track of the values of the digits.
- Have students model each step of U.S. traditional multiplication by using the money method simultaneously.
- Have students ignore the decimal points and solve the problem. Then they can use estimation to place the decimal point in the answer.

ELL Support

To support using estimation to place the decimal point in a product, have students consider a problem such as 2.9 * 4.2. Guide students to conclude that 2.9 is about 3, and 4.2 is about 4, so the answer should be about 3 * 4 = 12. When the decimal point is placed in 1218 (the product obtained when the decimal points are ignored), students conclude that 2.9 * 4.2 = 12.18. Encourage students to describe their thinking aloud as they estimate and place the decimal point in the product.

Writing/Reasoning

Have students write a response to the following: How is multiplying with decimals different from multiplying with whole numbers? *(Sample answer: Multiplying with decimals is different from multiplying with whole numbers because you need to insert the decimal point after you get the answer. You can count the total number of places to the right of the decimal point in each number you multiplied and then count over from the right that many places in the answer to place the decimal point.)*

Algorithm Project 7, U.S. Traditional Long Division, along with all other Grade 5 projects, is located at the back of both Volume 1 and Volume 2 of the Grade 5 *Teacher's Lesson Guide.* Use this project after Lesson 4-2.

Adjusting the Activity Ideas

- Encourage students to think of these problems in terms of fair sharing. Allow them to use manipulatives, like pencils in boxes or money, to illustrate this method.
- Have students use a place-value chart when using this method to keep track of the values of the digits.
- Have students use base-10 blocks to illustrate each step of U.S. traditional division. Use base-10 blocks either before or after recording the steps on paper.

ELL Support

Have students mark the appropriate columns with the place-value terms ones, tens, hundreds, and thousands. This will help them to remember the values of the digits they are dividing.

Writing/Reasoning

Have students write a response to the following: How would you explain this way of dividing to someone else? *(Sample answer: It is similar to partial quotients, but instead of writing each partial quotient to the right of the problem, you write the quotients above the appropriate digit in the dividend. You still write each product underneath the dividend and subtract. You might have to regroup. You continue dividing until you cannot divide anymore.)*

Algorithm
8
Project

U.S. Traditional Long Division with Decimal Dividends

Objective To extend long division to problems in which a decimal is divided by a whole number.

Algorithm Project 8, U.S. Traditional Long Division with Decimal Dividends, along with all other Grade 5 projects, is located at the back of both Volume 1 and Volume 2 of the Grade 5 *Teacher's Lesson Guide*. Use this project after Lesson 4-5 and Algorithm Project 7.

Adjusting the Activity Ideas

- Have students use a place-value chart when using this method to keep track of the values of the digits.
- Have students model each step of U.S. traditional division by using the money method simultaneously.
- Have students ignore the decimal points and solve the problem. Then they can use estimation to place the decimal point in the answer.

ELL Support

Have students mark the appropriate columns with the place-value terms ones, tens, hundreds, and, to the right of the decimal point, tenths and hundredths. This will help them to remember the values of the digits they are dividing.

Writing/Reasoning

Have students write a response to the following: How is dividing with decimals different from dividing with whole numbers? *(Sample answer: Dividing with decimals is different from dividing with whole numbers because you need to place the decimal point after you get the answer. You can place the decimal point directly above the decimal point in the dividend.)*

Algorithm
9
Project

U.S. Traditional Long Division: Decimals

Objective To extend long division to problems in which both the divisor and the dividend are decimals.

Algorithm Project 9, U.S. Traditional Long Division: Decimals, along with all other Grade 5 projects, is located at the back of both Volume 1 and Volume 2 of the Grade 5 *Teacher's Lesson Guide.* Use this project after Lesson 4-6 and Algorithm Project 8.

Adjusting the Activity Ideas

- Have students use a place-value chart when using this method to keep track of the values of the digits.
- Have students solve a simple division problem, such as 12 / 4. Then pose an equivalent problem, such as 120 / 40, and have students solve it. Point out that the answers are the same because they are equivalent problems.
- Have students ignore the decimal points and solve the problem. Then they can use estimation to place the decimal point in the answer.

ELL Support

Have students mark the appropriate columns with the place-value terms ones, tens, hundreds, and, to the right of the decimal point, tenths and hundredths. This will help them to remember the values of the digits they are dividing.

Writing/Reasoning

Have students write a response to the following: How is dividing with a decimal divisor different from dividing with a whole-number divisor? *(Sample answer: If the divisor is a decimal, you first have to change the problem to make the divisor a whole number. Then you divide like you do with whole numbers and place the decimal point in the answer.)*

Activities and Ideas for Differentiation

This section highlights Part 1 activities that support differentiation, optional Part 3 Readiness, Enrichment, Extra Practice, and ELL Support activities built into the lessons of the Grade 5 *Teacher's Lesson Guide,* and specific ideas for vocabulary development and games modifications. Provided in each unit is a chart showing where the Grade-Level Goals emphasized in that unit are addressed throughout the year. Following the chart, there are suggestions for maintaining concepts and skills to ensure that students continue working toward those Grade-Level Goals.

Contents

Unit 1 Activities and Ideas for Differentiation

In this unit, students explore finding factors and products and identifying prime and composite numbers. This section summarizes opportunities for supporting multiple learning styles and ability levels. Use these suggestions to develop a differentiation plan for Unit 1.

Part 1 Activities That Support Differentiation

Below are examples of Unit 1 activities that highlight some of the general instructional strategies that are hallmarks of a differentiated classroom. These strategies will help you support, emphasize, and enhance lesson content to make sure all of your students are engaged in the mathematics at the highest possible level. For more information about general differentiation strategies that accommodate the diverse needs of today's classrooms, see the essay on pages 8–16 of this handbook.

Lesson	Activity	Strategy
1•2	**Students draw rectangular arrays for numbers.**	Modeling visually
1•4	**Students discuss the meaning of divisibility.**	Talking about math
1•6	**Students record factor rainbows for numbers.**	Using organizational tools
1•7	**Students draw arrays to represent square numbers.**	Modeling visually
1•8	**Record information about square roots on the board.**	Recording key ideas
1•9	**Students make factor trees to find factors of numbers.**	Using organizational tools

Vocabulary Development

The list below identifies the Key Vocabulary terms from this unit. The lesson in which each term is defined is indicated next to the term. Some of these terms or their homophones are used outside of mathematics. Consider adding other words as appropriate for developing understanding of the context of the lessons.

Lessons include suggestions for helping English language learners understand and develop vocabulary. For more information, see pages 17–19 of this handbook.

Key Vocabulary

Commutative Property of Multiplication **1◆2**

composite number **1◆6**

divisibility rule (*rule) **1◆5**

divisible by (†by) **1◆4**

even number **1◆4**

exponent **1◆7**

exponent key (*†key) **1◆7**

exponential notation **1◆7**

*factor **1◆3**

factor pair (†pair) **1◆3**

factor rainbow **1◆5**

factor string (*string) **1◆9**

length of factor string **1◆9**

name-collection box (*box) **1◆9**

number model (*model) **1◆2**

odd number (*odd) **1◆4**

prime factorization (*prime) **1◆9**

prime number **1◆6**

*product **1◆3**

quotient **1◆5**

rectangular array **1◆2**

remainder **1◆4**

square array **1◆7**

square number **1◆7**

square root (*†root) **1◆8**

square-root key **1◆8**

turn-around rule (for multiplication) (†turn) **1◆2**

unsquaring a number **1◆8**

* Discuss the everyday and mathematical meanings of the words that are marked with an asterisk.

† For words marked with a dagger, write the words and their homophones on the board. For example, *by, buy,* and *bye; key* and *quay; pair, pare,* and *pear; root* and *route;* and *turn* and *tern.* Discuss and clarify the meaning of each.

◆ As each word is introduced in the lesson, write the word on the board and discuss its meaning.

◆ List the words on a Math Word Wall for students to see. As each word is introduced in the lesson, add a picture next to the word on the Word Wall.

◆ Use the vocabulary words regularly when teaching lessons, and encourage students to use the words in their discussions.

Games

Below are suggested Unit 1 game adaptations. For more information about implementing games in a differentiated classroom, see pages 20–25 of this handbook.

Game: *Multiplication Top-It* (Extended-Facts Version)

Skill Practiced: Solve multiplication fact extensions. [Operations and Computation Goal 2]

Modification	Purpose of Modification
Players keep one factor constant and turn over a card for the second factor in each round. If 5 is the constant, players draw a card and multiply that number by 5.	Students solve a group of multiplication fact extensions with one factor in common. [Operations and Computation Goal 2]
Players use the number cards 1–20. They turn over two cards, attach a zero to each card, and multiply.	Students solve multidigit multiplication problems. [Operations and Computation Goal 2]

Game: *Factor Captor*

Skill Practiced: Identify factors of numbers. [Number and Numeration Goal 3; Operations and Computation Goal 2]

Modification	Purpose of Modification
Before playing the game, players use a different color to shade each of the 2-digit numbers from the *Factor Captor* Grid on a Multiplication/Division Facts Table every place they appear. They use the table as a reference when they play the game.	Students identify factors of a number using a Multiplication/Division Facts Table. [Number and Numeration Goal 3; Operations and Computation Goal 2]
Players use a number grid as the gameboard. For each number that is covered first, players make a factor tree. They get a bonus point for each correct tree.	Students identify factors of numbers. [Number and Numeration Goal 3]

Game: *Factor Bingo*

Skill Practiced: Identify factors of numbers. [Number and Numeration Goal 3]

Modification	Purpose of Modification
Players fill their gameboard with multiples of 5 from 0 to 100. (They may use a number two times.) They play with four each of the number cards 0–10 and one each of 11–20. They draw one card on each turn. Players cover one square on their gameboard for each turn.	Students identify factors of numbers that are multiples of 5. [Number and Numeration Goal 3]
Players play on a number grid. They get a bingo for any six numbers in a row.	Students identify factors of numbers. [Number and Numeration Goal 3]

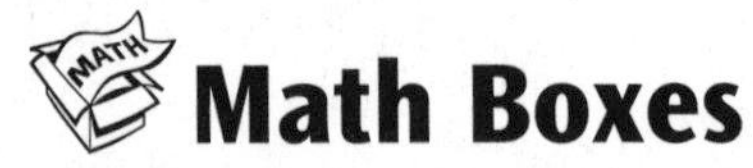

Math Boxes

Suggestions for using Math Boxes to meet individual needs begin on page 26 of this handbook. There are blank masters for Math Boxes on pages 136–141.

Using Part 3 of the Lessons

Use your professional judgment, along with assessment results, to determine whether the whole class, small groups, or individual students might benefit from these Unit 1 activities. Consider using the Part 3 Planning Master found on page 154 of this handbook to record your plans.

Readiness Activities

Lesson	Activity	Purpose of Activity
1•1	**Follow a set of directions.**	Explore a variety of problem-solving situations. [Number and Numeration Goal 1]
1•2	**Use counters to build arrays and find the total number of counters in each.**	Explore factoring numbers. [Number and Numeration Goal 3]
1•3	**Build arrays using centimeter cubes to highlight the relationships between arrays and factor pairs.**	Gain experience finding factor pairs. [Number and Numeration Goal 3]
1•4	**Explore multiplication and division relationships using fact families.**	Explore the inverse relationship between multiplication and division. [Patterns, Functions, and Algebra Goal 4]
1•5	**Use counters to determine whether a number is divisible by the numbers 1–6.**	Explore the concept of divisibility. [Operations and Computation Goal 3]
1•7	**Build arrays for products that are square numbers.**	Explore the relationship between the doubles multiplication facts and square numbers. [Operations and Computation Goal 7]
1•9	**Use the Sieve of Eratosthenes to find the prime numbers.**	Explore the concept of prime numbers. [Number and Numeration Goal 3]

English Language Learners Support Activities

Lesson	Activity	Purpose of Activity
1•2	**Describe *arrays, rows,* and *columns* in the Arrays Museum.**	Make connections between mathematics and everyday life; discuss new mathematical ideas. [Operations and Computation Goal 7]
1•3	**Add *factor, product,* and *multiplication* to the Math Word Bank.**	Make connections among and use visuals to represent terms. [Operations and Computation Goal 3]
1•5	**Add *divisor, dividend, quotient,* and *remainder* to the Math Word Bank.**	Make connections among and use visuals to represent terms. [Operations and Computation Goal 3]

Enrichment Activities

Lesson	Activity	Purpose of Activity
1•2	**Solve and create magic square and heterosquare arrays.**	Explore rectangular arrays. [Operations and Computation Goal 7]
1•4	**Play *Factor Captor.***	Apply strategies for finding factors of larger numbers. [Number and Numeration Goal 3]
1•5	**Use place-value concepts to explore a test for divisibility by 4.**	Explore divisibility. [Operations and Computation Goal 3]
1•6	**Explore Goldbach's Conjecture.**	Explore prime numbers. [Number and Numeration Goal 3]
1•7	**Draw, label, and describe dot patterns.**	Explore dot patterns in rectangular and nonrectangular arrays. [Patterns, Functions, and Algebra Goal 1]
1•8	**Investigate the relationship between numbers and their squares.**	Explore factoring numbers. [Number and Numeration Goal 3]
1•9	**Find palindromic numbers and squares.**	Explore the relationship between numbers and their squares. [Number and Numeration Goal 3]

Extra Practice Activities

Lesson	Activity	Purpose of Activity
1•1	**Solve *5-Minute Math* problems involving Roman numerals.**	Practice with different representations of numbers. [Number and Numeration Goal 4]
1•3	**Solve *5-Minute Math* problems involving factoring whole numbers.**	Practice factoring whole numbers. [Number and Numeration Goal 3]
1•3	**Use a multiplication facts routine.**	Practice multiplication facts. [Operations and Computation Goal 2]
1•5	**Use a multiplication facts routine.**	Practice multiplication facts. [Operations and Computation Goal 2]
1•6	**Solve *5-Minute Math* problems involving square roots.**	Practice with the square-root symbol. [Number and Numeration Goal 3]
1•8	**Solve *5-Minute Math* problems involving square roots.**	Practice with the square-root symbol. [Number and Numeration Goal 3]
1•9	**Use a multiplication facts routine.**	Practice multiplication facts. [Operations and Computation Goal 2]

Looking at Grade-Level Goals

Everyday Mathematics develops concepts and skills over time. Below is a chart showing where the Grade-Level Goals emphasized in this unit are addressed throughout the year. Use the chart to help you determine which Maintaining Concepts and Skills activities on page 56 to utilize to ensure that students continue working toward these Grade-Level Goals.

● Grade-Level Goal is taught.
◍ Grade-Level Goal is practiced and applied.
○ Grade-Level Goal is not a focus.

Grade-Level Goals Emphasized in Unit 1	Unit											
	1	2	3	4	5	6	7	8	9	10	11	12
[Number and Numeration Goal 3] Identify prime and composite numbers; factor numbers; find prime factorizations.	●	◍	◍	◍	◍	○	○	○	◍	○	◍	◍
[Number and Numeration Goal 4] Use numerical expressions involving one or more of the basic four arithmetic operations, grouping symbols, and exponents to give equivalent names for whole numbers; convert between base-10, exponential, and repeated-factor notations.	●	◍	◍	○	○	○	●	○	○	◍	◍	○
[Operations and Computation Goal 2] Demonstrate automaticity with multiplication and division fact extensions.	●	◍	◍	●	○	○	○	○	○	○	○	○
[Operations and Computation Goal 3] Use manipulatives, mental arithmetic, paper-and-pencil algorithms and models, and calculators to solve problems involving the multiplication of whole numbers and decimals and the division of multidigit whole numbers and decimals by whole numbers; express remainders as whole numbers or fractions as appropriate; describe the strategies used and explain how they work.	●	●	◍	●	◍	◍	◍	◍	◍	◍	◍	◍
[Operations and Computation Goal 7] Use repeated addition, arrays, area, and scaling to model multiplication and division; use ratios expressed as words, fractions, percents, and with colons; solve problems involving ratios of parts of a set to the whole set.	●	◍	○	○	◍	○	◍	●	◍	●	○	◍
[Patterns, Functions, and Algebra Goal 1] Extend, describe, and create numeric patterns; describe rules for patterns and use them to solve problems; write rules for functions involving the four basic arithmetic operations; represent functions using words, symbols, tables, and graphs and use those representations to solve problems.	●	○	◍	○	○	○	●	●	◍	●	○	◍
[Patterns, Functions, and Algebra Goal 4] Describe and apply properties of arithmetic.	●	◍	○	○	○	○	◍	○	○	○	◍	○

Maintaining Concepts and Skills

Many of the goals addressed in this unit will be addressed again in later units. Those goals marked with an asterisk (*) are addressed in future units only as practice and application. Here are several suggestions for maintaining concepts and skills until goals are revisited.

Number and Numeration Goal 4

- Have students play *Name That Number.*
- Use the Name-Collection Boxes master on page 147 of this handbook to create practice problems.

Operations and Computation Goal 2

- Have students play *Multiplication Top-It* (Extended-Facts Version).
- Have students practice with Fact Triangles.

Operations and Computation Goal 7

- Have students build arrays and find the total number of counters for arrays. See the Readiness activity in Lesson 1-2 for more information.
- Have students investigate square-number facts. See the Readiness activity in Lesson 1-7 for more information.

Patterns, Functions, and Algebra Goal 4*

- Have students explore relationships between multiplication and division. See the Readiness activity in Lesson 1-4 for more information.
- Have students practice with Fact Triangles and have them record fact families for Fact Triangles.

Assessment

See page 54 in the *Assessment Handbook* for modifications to the written portion of the Unit 1 Progress Check.

Additionally, see pages 55–59 for modifications to the open-response task and selected student work samples.

Unit 2 Activities and Ideas for Differentiation

In this unit, students extend their work with estimating answers to computation problems, pencil-and-paper computation algorithms, and data collection. This section summarizes opportunities for supporting multiple learning styles and ability levels. Use these suggestions to develop a differentiation plan for Unit 2.

Part 1 Activities That Support Differentiation

Below are examples of Unit 2 activities that highlight some of the general instructional strategies that are hallmarks of a differentiated classroom. These strategies will help you support, emphasize, and enhance lesson content to make sure all of your students are engaged in the mathematics at the highest possible level. For more information about general differentiation strategies that accommodate the diverse needs of today's classrooms, see the essay on pages 8–16 of this handbook.

Lesson	Activity	Strategy
2•2	**Students use partial-sums and column-addition methods to solve problems.**	Incorporating and validating a variety of methods
2•4	**Record information about number sentences on the board.**	Recording key ideas
2•5	**Students make stick-on note line plots for a reaction-times experiment.**	Modeling concretely
2•6	**Students use the Probability Meter as a reference for fraction, decimal, and percent names for probabilities.**	Using a visual reference
2•8	**Students review the partial-products method for whole-number multiplication.**	Building on prior knowledge
2•10	**Students make posters and compare strategies for finding the number of minutes in a year.**	Talking about math

Vocabulary Development

The list below identifies the Key Vocabulary terms from this unit. The lesson in which each term is defined is indicated next to the term. Some of these terms or their homophones are used outside of mathematics. Consider adding other words as appropriate for developing understanding of the context of the lessons.

Lessons include suggestions for helping English language learners understand and develop vocabulary. For more information, see pages 17–19 of this handbook.

Key Vocabulary

algorithm **2♦2**
ballpark estimate **2♦8**
certain **2♦6**
column-addition method (*column) **2♦2**
*difference **2♦3**
*digit **2♦2**
estimate **2♦1**
expanded notation **2♦2**
false number sentence **2♦4**
impossible **2♦6**
*lattice **2♦9**
lattice method **2♦9**
magnitude estimate **2♦7**
maximum **2♦5**
*†mean (average) **2♦5**
*median **2♦1**
minimum **2♦5**
minuend **2♦3**
*†mode **2♦5**
number sentence **2♦4**
open number sentence (*open) **2♦4**
operation symbol (†symbol) **2♦4**
partial-differences method **2♦3**
partial-products method **2♦8**
partial-sums method (†sum) **2♦2**
*†place **2♦2**
place value **2♦2**
Probability Meter Poster **2♦6**
*range **2♦5**
reaction time (†time) **2♦5**
relation symbol (*relation) **2♦4**
sample **2♦10**
*solution **2♦4**
stimulus **2♦5**
subtrahend **2♦3**
trade-first method **2♦3**
true number sentence **2♦4**
value **2♦2**
variable **2♦4**

* Discuss the everyday and mathematical meanings of the words that are marked with an asterisk.

† For words marked with a dagger, write the words and their homophones on the board. For example, *mean* and *mien; mode* and *mowed; symbol* and *cymbal; sum* and *some; place* and *plaice;* and *time* and *thyme.* Discuss and clarify the meaning of each.

◆ As each word is introduced in the lesson, write the word on the board and discuss its meaning.

◆ List the words on a Math Word Wall for students to see. As each word is introduced in the lesson, add a picture next to the word on the Word Wall.

◆ Use the vocabulary words regularly when teaching lessons, and encourage students to use the words in their discussions.

Games

Below are suggested Unit 2 game adaptations. For more information about implementing games in a differentiated classroom, see pages 20–25 of this handbook.

Game: *Addition Top-It* (Decimal Version)	
Skill Practiced: Solve addition problems. [Operations and Computation Goal 1]	
Modification	**Purpose of Modification**
Players keep one addend constant and turn over the second addend for each round. If the constant is 5, players draw one card and add that number to 5.	Students solve addition problems with one constant addend. [Operations and Computation Goal 1]
Players use the number cards 1–20. They turn over six cards and find the sum of three decimal numbers.	Students solve addition problems with decimals. [Operations and Computation Goal 1]

Game: *Subtraction Target Practice*	
Skill Practiced: Solve subtraction problems. [Operations and Computation 1]	
Modification	**Purpose of Modification**
Players turn over one card on each turn. They subtract that number or that many tens from their total. If a player turns over a 3, either 3 or 30 can be subtracted.	Students solve subtraction problems. [Operations and Computation 1]
Players draw two cards and either subtract a 2-digit number that they make with the cards or subtract the sum of the two cards from their total.	Students solve addition and subtraction problems. [Operations and Computation 1]

Game: *High-Number Toss*	
Skill Practiced: Read, write, and compare numbers. [Number and Numeration Goals 1 and 6]	
Modification	**Purpose of Modification**
Players use all four spaces as digits so each number they build is in the thousands.	Students read, write, and compare numbers in the thousands. [Number and Numeration Goals 1 and 6]
After all of the spaces are filled in, players can make one "switch" trading the digits in two of the places. Then they record and compare their numbers.	Students read, write, and compare numbers. [Number and Numeration Goals 1 and 6]

Math Boxes

Suggestions for using Math Boxes to meet individual needs begin on page 26 of this handbook. There are blank masters for Math Boxes on pages 136–141.

Using Part 3 of the Lessons

Use your professional judgment, along with assessment results, to determine whether the whole class, small groups, or individual students might benefit from these Unit 2 activities. Consider using the Part 3 Planning Master found on page 154 of this handbook to record your plans.

Readiness Activities

Lesson	Activity	Purpose of Activity
2•1	**Estimate the total number of objects in a visual field.**	Gain experience developing estimation strategies. [Operations and Computation Goal 6]
2•2	**Build numbers with base-10 blocks and use the blocks to add.**	Explore place value and expanded notation. [Number and Numeration Goal 1]
2•3	**Model subtraction by making and breaking apart bundles of ten sticks or straws.**	Gain experience trading between place-value columns. [Operations and Computation Goal 1]
2•4	**Use situation diagrams to write open number sentences.**	Gain experience using situation diagrams. [Patterns, Functions, and Algebra Goal 2]
2•5	**Skip count on a calculator to find missing decimals on a number line.**	Explore ordering decimals. [Number and Numeration Goal 6]
2•6	**Use number cards to order fractions, decimals, and percents.**	Explore ordering fractions, decimals, and percents. [Number and Numeration Goal 6]
2•7	**Use multiplication basic facts and patterns to solve extended-fact problems.**	Explore multiplication patterns. [Operations and Computation Goal 2; Patterns, Functions, and Algebra Goal 1]
2•8	**Model the partial-products multiplication method using base-10 blocks.**	Gain experience with multiplication. [Operations and Computation Goal 3]
2•10	**Play *Number Top-It*.**	Gain experience with place-value concepts. [Number and Numeration Goals 1 and 6]

English Language Learners Support Activities

Lesson	Activity	Purpose of Activity
2•2	**Add *expanded notation* to the Math Word Bank.**	Make connections among and use visuals to represent terms. [Number and Numeration Goal 1]

Enrichment Activities

Lesson	Activity	Purpose of Activity
2•1	**Develop a strategy for estimating the number of names on a phone-book page.**	Apply understanding of estimation skills. [Operations and Computation Goal 6]
2•2	**Use the sum of one problem to find the solution of other problems.**	Apply understanding of place value and addition algorithms. [Operations and Computation Goal 1]
2•3	**Compare different subtraction methods.**	Explore subtraction algorithms. [Operations and Computation Goal 1]
2•4	**Solve multistep number stories.**	Apply understanding of open number sentences. [Patterns, Functions, and Algebra Goal 2]
2•5	**Reorganize collected data and interpret the data based on the reorganization.**	Explore organizing and analyzing data. [Data and Chance Goals 1 and 2]
2•6	**Design spinners to match given situations.**	Apply understanding of probability concepts and explore the language of probability. [Data and Chance Goals 3 and 4]
2•7	**Use patterns in the number of zeros in factors and products.**	Apply understanding of multiplying by powers of 10. [Number and Numeration Goal 1; Operations and Computation Goal 2]
2•8	**Devise a mental-multiplication strategy for multiplying by 9.**	Explore multiplication strategies [Operations and Computation Goal 3]
2•9	**Explore and analyze how to use an ancient Egyptian multiplication method.**	Apply understanding of multiplication. [Operations and Computation Goal 3]
2•10	**Write number sentences and make estimates for number stories.**	Apply estimation strategies to solving problems. [Operations and Computation Goal 6; Patterns, Functions, and Algebra Goal 2]

Extra Practice Activities

Lesson	Activity	Purpose of Activity
2•7	**Solve *5-Minute Math* problems involving estimation.**	Practice estimating. [Operations and Computation Goal 6]
2•9	**Solve *5-Minute Math* problems involving multiplying decimals.**	Practice multiplying decimals. [Operations and Computation Goal 3]
2•9	**Solve problems involving multiplying decimals.**	Practice multiplying decimals. [Operations and Computation Goal 3]
2•10	**Solve problems involving place value with powers of 10.**	Gain more experience with place value. [Number and Numeration Goal 1]

Looking at Grade-Level Goals

Everyday Mathematics develops concepts and skills over time. Below is a chart showing where the Grade-Level Goals emphasized in this unit are addressed throughout the year. Use the chart to help you determine which Maintaining Concepts and Skills activities on page 63 to utilize to ensure that students continue working toward these Grade-Level Goals.

● Grade-Level Goal is taught.
◎ Grade-Level Goal is practiced and applied.
○ Grade-Level Goal is not a focus.

Grade-Level Goals Emphasized in Unit 2	Unit											
	1	2	3	4	5	6	7	8	9	10	11	12
[Number and Numeration Goal 1] Read and write whole numbers and decimals; identify places in such numbers and the values of the digits in those places; use expanded notation to represent whole numbers and decimals.	◎	●	◎	◎	◎	◎	●	◎	◎	◎	○	○
[Operations and Computation Goal 1] Use manipulatives, mental arithmetic, paper-and-pencil algorithms and models, and calculators to solve problems involving the addition and subtraction of whole numbers, decimals, and signed numbers; describe the strategies used and explain how they work.	◎	●	◎	◎	◎	◎	●	◎	◎	◎	◎	○
[Operations and Computation Goal 3] Use manipulatives, mental arithmetic, paper-and-pencil algorithms and models, and calculators to solve problems involving the multiplication of whole numbers and decimals and the division of multidigit whole numbers and decimals by whole numbers; express remainders as whole numbers or fractions as appropriate; describe the strategies used and explain how they work.	●	●	◎	●	◎	◎	◎	◎	◎	◎	◎	◎
[Operations and Computation Goal 6] Make reasonable estimates for whole number and decimal addition, subtraction, multiplication, and division problems and fraction and mixed number addition and subtraction problems; explain how the estimates were obtained.	◎	●	◎	●	○	◎	○	◎	○	◎	○	○
[Data and Chance Goal 1] Collect and organize data or use given data to create graphic displays with reasonable titles, labels, keys, and intervals.	○	◎	○	○	●	●	◎	◎	○	◎	◎	○
[Data and Chance Goal 2] Use the maximum, minimum, range, median, mode, and mean and graphs to ask and answer questions, draw conclusions, and make predictions.	◎	●	●	◎	◎	●	◎	○	◎	◎	○	○
[Data and Chance Goal 3] Describe events using *certain, very likely, likely, unlikely, very unlikely, impossible,* and other basic probability terms; use *more likely, equally likely, same chance, 50-50, less likely,* and other basic probability terms to compare events; explain the choice of language.	◎	●	○	○	◎	○	◎	○	◎	○	○	◎
[Data and Chance Goal 4] Predict the outcomes of experiments, test the predictions using manipulatives, and summarize the results; compare predictions based on theoretical probability with experimental results; use summaries and comparisons to predict future events; express the probability of an event as a fraction, decimal, or percent.	◎	●	○	○	◎	●	◎	○	◎	○	○	◎
[Patterns, Functions, and Algebra Goal 2] Determine whether number sentences are true or false; solve open number sentences and explain the solutions; use a letter variable to write an open sentence to model a number story; use a pan-balance model to solve linear equations in one unknown.	○	◎	◎	●	◎	◎	●	◎	○	●	◎	◎

Maintaining Concepts and Skills

Many of the goals addressed in this unit will be addressed again in later units. Those goals marked with an asterisk (*) are addressed in future units only as practice and application. Here are several suggestions for maintaining concepts and skills until goals are revisited.

Number and Numeration Goal 1

- Have students play *Number Top-It* and *High-Number Toss.*

Operations and Computation Goal 3

- Have students use base-10 blocks to make an area model for the partial-products algorithm. See the Readiness activity in Lesson 2-8 for more information.
- Use the "What's My Rule?" master on page 146 of this handbook to create practice problems with multiplication and division rules.

Operations and Computation Goal 6

- Have students play *Multiplication Bull's-Eye.*
- Have students make estimates. See the Readiness activity in Lesson 2-1 for more information.

Data and Chance Goal 3*

- Have students routinely post probabilities for events on the Probability Meter, for example, the probability that everyone will turn in his or her homework, that it will rain tomorrow, or that there will be outdoor recess.
- Have students compare probabilities using the language of chance.

Assessment

See page 62 in the *Assessment Handbook* for modifications to the written portion of the Unit 2 Progress Check.

Additionally, see pages 63–67 for modifications to the open-response task and selected student work samples.

Unit 3 Activities and Ideas for Differentiation

In this unit, students will be introduced to the American Tour and further explore plane figures and tessellations. This section summarizes opportunities for supporting multiple learning styles and ability levels. Use these suggestions to develop a differentiation plan for Unit 3.

Part 1 Activities That Support Differentiation

Below are examples of Unit 3 activities that highlight some of the general instructional strategies that are hallmarks of a differentiated classroom. These strategies will help you support, emphasize, and enhance lesson content to make sure all of your students are engaged in the mathematics at the highest possible level. For more information about general differentiation strategies that accommodate the diverse needs of today's classrooms, see the essay on pages 8–16 of this handbook.

Lesson	Activity	Strategy
3◆1	**Students post U.S. census percentages on the Probability Meter.**	Using a visual reference
3◆3	**Students use pattern blocks to find angle measures in polygons.**	Modeling concretely
3◆4	**Students use their arms to demonstrate angles of different types.**	Modeling physically
3◆6	**Students share their strategies for copying a triangle.**	Incorporating and validating a variety of methods
3◆8	**Students use paper polygons to explore tessellations.**	Modeling concretely
3◆9	**Students discuss and describe interior angles in polygons.**	Talking about math

Vocabulary Development

The list below identifies the Key Vocabulary terms from this unit. The lesson in which each term is defined is indicated next to the term. Some of these terms or their homophones are used outside of mathematics. Consider adding other words as appropriate for developing understanding of the context of the lessons.

Lessons include suggestions for helping English language learners understand and develop vocabulary. For more information, see pages 17–19 of this handbook.

Key Vocabulary

acute angle (*acute) **3◆4**
adjacent angles **3◆5**
†arc **3◆4**
†census **3◆1**
congruent **3◆6**
diameter **3◆5**
equilateral triangle **3◆6**
Geometry Template **3◆4**
isosceles triangle **3◆6**
obtuse angle (*obtuse) **3◆4**
pentagon **3◆10**
perimeter **3◆10**
polygon **3◆7**
quadrangle **3◆7**
radius **3◆5**
reflex angle (*reflex) **3◆4**
regular polygon (*regular) **3◆8**
regular tessellation **3◆8**
right angle (*†right) **3◆4**
scalene triangle **3◆6**
straight angle (†straight) **3◆4**
tessellate **3◆8**
tessellation **3◆8**
tessellation vertex **3◆8**
vertical (or opposite) angles **3◆5**

* Discuss the everyday and mathematical meanings of the words that are marked with an asterisk.

† For words marked with a dagger, write the words and their homophones on the board. For example, *arc* and *ark; census* and *senses; right, write, rite,* and *wright;* and *straight* and *strait.* Discuss and clarify the meaning of each.

◆ As each word is introduced in the lesson, write the word on the board and discuss its meaning.

◆ List the words on a Math Word Wall for students to see. As each word is introduced in the lesson, add a picture next to the word on the Word Wall.

◆ Use the vocabulary words regularly when teaching lessons, and encourage students to use the words in their discussions.

Games

Below are suggested Unit 3 game adaptations. For more information about implementing games in a differentiated classroom, see pages 20–25 of this handbook.

Game: *High-Number Toss*—Decimal Version	
Skill Practiced: Identify place value and compare decimal numbers. [Number and Numeration Goals 1 and 6]	
Modification	**Purpose of Modification**
Players move the decimal point so that the number they build has a whole number and two decimal places.	Students read, write, and compare numbers in a specified place value. [Number and Numeration Goals 1 and 6]
Players play the decimal version like the whole-number version of *High-Number Toss*. They use the number on the last blank to tell how many decimal places to move the decimal point to the left from the vertical dash. The final number could be as large as in the tens and as small as the millionths.	Students read, write, and compare decimal numbers. [Number and Numeration Goals 1 and 6]

Game: *Angle Tangle*	
Skill Practiced: Estimate the size of and measure angles. [Measurement and Reference Frames Goal 1]	
Modification	**Purpose of Modification**
Players estimate whether angles are larger or smaller than a right angle (90°). They can use the corner of a piece of paper to check their estimates.	Students compare angles to 90° angles and measure angles. [Measurement and Reference Frames Goal 1]
Each player draws an angle. Both players estimate the difference between the two angle measures. The player with the closer estimate scores a point. The winner is the player with the most points.	Students estimate the size of and measure angles. [Measurement and Reference Frames Goal 1]

Game: *Polygon Capture*	
Skill Practiced: Describe properties of polygons. [Geometry Goal 2]	
Modification	**Purpose of Modification**
Players use only the Property Cards. They draw one Property Card and use their Geometry Templates to draw a shape that has the property. They receive one point if they correctly draw a shape.	Students draw polygons according to polygon properties. [Geometry Goal 2]
Using two sets of Property Cards and one set of Polygon Capture Pieces, players draw two Property Cards on each turn. They can use one or two Property Cards to make a match collecting only one Polygon Capture Piece on a turn. The player with the most cards at the end wins.	Students describe properties of polygons. [Geometry Goal 2]

Math Boxes

Suggestions for using Math Boxes to meet individual needs begin on page 26 of this handbook. There are blank masters for Math Boxes on pages 136–141.

Using Part 3 of the Lessons

Use your professional judgment, along with assessment results, to determine whether the whole class, small groups, or individual students might benefit from these Unit 3 activities. Consider using the Part 3 Planning Master found on page 154 of this handbook to record your plans.

Readiness Activities

Lesson	Activity	Purpose of Activity
3•1	**Read text, tables, charts, and graphs for information and answer questions based on data.**	Explore analyzing data. [Data and Chance Goal 2]
3•2	**Write large numbers in place-value charts and solve place-value riddles.**	Explore place value. [Number and Numeration Goal 1]
3•3	**Draw and name an angle in different ways.**	Explore naming angles. [Geometry Goal 1]
3•4	**Solve riddles about points, lines, line segments, and angles.**	Gain experience with vocabulary and concepts related to angles. [Geometry Goal 1]
3•5	**Find equivalent measures and measure line segments in different ways.**	Explore measuring with a ruler. [Measurement and Reference Frames Goal 1]
3•6	**Play *Triangle Sort.***	Explore properties of triangles. [Geometry Goal 2]
3•7	**Sort attribute blocks by two properties at a time.**	Gain experience with geometric properties. [Geometry Goal 2]
3•8	**Use pattern blocks to make tessellations.**	Explore tessellations. [Geometry Goal 3]
3•10	**Make posters to illustrate geometry terms.**	Create visual references for geometry vocabulary. [Geometry Goals 1 and 2]

English Language Learners Support Activities

Lesson	Activity	Purpose of Activity
3•4	**Add *acute angle, right angle, obtuse angle, straight angle,* and *reflex angle* to the Math Word Bank.**	Make connections among and use visuals to represent terms. [Geometry Goal 1]
3•5	**Add *diameter* and *radius* to the Math Word Bank.**	Make connections among and use visuals to represent terms. [Measurement and Reference Frames Goal 2]
3•7	**Play *What's My Attribute Rule?* and discuss the attribute rules from the game.**	Clarify the mathematical uses of the term. [Patterns, Functions, and Algebra Goal 1]
3•9	**Describe *tessellations, polygons, colors, shapes,* and *patterns* in the Tessellation Museum.**	Discuss new mathematical ideas. [Geometry Goal 3]

Enrichment Activities

Lesson	Activity	Purpose of Activity
3•1	**Answer questions based on a table of data taken from the U.S. Census.**	Explore analyzing data. [Data and Chance Goal 2]
3•2	**Analyze patterns in data displayed on a map.**	Apply understanding of data. [Data and Chance Goal 2]
3•3	**Use clues to name line segments and collinear points.**	Explore naming points and line segments. [Geometry Goal 1]
3•4	**Solve problems that involve baseball-field angles.**	Apply understanding of angle properties and measures. [Measurement and Reference Frames Goal 1]
3•5	**Inscribe a regular hexagon in a circle.**	Explore straightedge constructions. [Geometry Goal 2]
3•6	**Play *Sides and Angles: Triangles.***	Apply understanding of the relationships among the sides and angles of triangles. [Geometry Goal 2]
3•7	**Connect the vertices of polygons to make new polygons and name the new polygons.**	Apply understanding of properties of polygons. [Geometry Goal 2]
3•8	**Use Geometry Template polygons to create and label tessellations.**	Explore tessellations. [Geometry Goal 3]
3•9	**Investigate whether all quadrangles will tessellate.**	Apply understanding of measures of interior angles of polygons. [Geometry Goal 3]
3•10	**Use the Geometry Template to solve problems involving polygons and angle relationships.**	Explore polygons. [Geometry Goal 2]

Extra Practice Activities

Lesson	Activity	Purpose of Activity
3•2	**Read bar graphs.**	Practice interpreting data. [Data and Chance Goal 2]
3•3	**Solve problems involving degree measures of angles in a circle.**	Practice with degree measures of congruent circle sectors. [Measurement and Reference Frames Goal 1]
3•6	**Play *Where Do I Fit In?***	Practice identifying side and angle properties for specific triangles. [Geometry Goal 2]
3•9	**Complete a table showing the relationship between the number of sides of a polygon and the number of interior triangles.**	Practice finding the number of interior triangles in a polygon. [Geometry Goal 2]
3•10	**Measure angles.**	Practice measuring angles. [Measurement and Reference Frames Goal 1]

Looking at Grade-Level Goals

Everyday Mathematics develops concepts and skills over time. Below is a chart showing where the Grade-Level Goals emphasized in this unit are addressed throughout the year. Use the chart to help you determine which Maintaining Concepts and Skills activities on page 70 to utilize to ensure that students continue working toward these Grade-Level Goals.

● Grade-Level Goal is taught.
◍ Grade-Level Goal is practiced and applied.
○ Grade-Level Goal is not a focus.

Grade-Level Goals Emphasized in Unit 3	Unit											
	1	2	3	4	5	6	7	8	9	10	11	12
[Data and Chance Goal 2] Use the maximum, minimum, range, median, mode, and mean and graphs to ask and answer questions, draw conclusions, and make predictions.	◍	●	●	◍	◍	●	◍	○	◍	◍	○	○
[Measurement and Reference Frames Goal 1] Estimate length with and without tools; measure length with tools to the nearest $\frac{1}{8}$ inch and millimeter; estimate the measure of angles with and without tools; use tools to draw angles with given measures.	○	◍	●	●	●	●	◍	◍	◍	○	○	○
[Geometry Goal 1] Identify, describe, compare, name, and draw right, acute, obtuse, straight, and reflex angles; determine angle measures in vertical and supplementary angles and by applying properties of sums of angle measures in triangles and quadrangles.	○	◍	●	◍	○	○	○	○	○	○	○	◍
[Geometry Goal 2] Describe, compare, and classify plane and solid figures using appropriate geometric terms; identify congruent figures and describe their properties.	○	◍	●	○	○	○	○	◍	◍	◍	◍	○

Maintaining Concepts and Skills

Some of the goals addressed in this unit will be addressed again in later units. Those goals marked with an asterisk (*) are addressed in future units only as practice and application. Here are several suggestions for maintaining concepts and skills until goals are revisited.

Data and Chance Goal 2

- Have students read tables, charts, and graphs. See the Readiness activity in Lesson 3-1 for more information.

Measurement and Reference Frames Goal 1

- Have students play *Angle Tangle.*
- Have students measure line segments. See the Readiness activity in Lesson 3-5 for more information.
- Use the Name-Collection Boxes master on page 147 of this handbook to create practice problems. Have students include measurement names for numbers; for example, for 12, one name is *the number of inches in a foot.*

Geometry Goal 1*

- Have students review angle names. See the Readiness activity in Lesson 3-3 for more information.
- Have students identify points, lines, and angles. See the Readiness activity in Lesson 3-4 for more information.
- Have students model parallel and intersecting line segments and angles of a variety of measures with their arms.

Geometry Goal 2*

- Have students play *Polygon Capture*.
- Have students sort attribute blocks according to properties. See the Readiness activity in Lesson 3-7 for more information.

Assessment

See page 70 in the *Assessment Handbook* for modifications to the written portion of the Unit 3 Progress Check.

Additionally, see pages 71–75 for modifications to the open-response task and selected student work samples.

Activities and Ideas for Differentiation

In this unit, students review division facts and the partial-quotients division algorithm, and they extend the algorithm to division of decimals. This section summarizes opportunities for supporting multiple learning styles and ability levels. Use these suggestions to develop a differentiation plan for Unit 4.

Part 1 Activities That Support Differentiation

Below are examples of Unit 4 activities that highlight some of the general instructional strategies that are hallmarks of a differentiated classroom. These strategies will help you support, emphasize, and enhance lesson content to make sure all of your students are engaged in the mathematics at the highest possible level. For more information about general differentiation strategies that accommodate the diverse needs of today's classrooms, see the essay on pages 8–16 of this handbook.

Lesson	Activity	Strategy
4•2	**Record the steps for the partial-quotients algorithm on the Class Data Pad.**	Using a visual reference
4•3	**Students solve map-scale problems.**	Connecting to everyday life
4•4	**Students use multiples to solve division problems.**	Building on prior knowledge
4•5	**Students share strategies for making estimates of quotients for decimal division.**	Incorporating and validating a variety of methods
4•6	**Students draw pictures to represent division problems.**	Modeling visually
4•7	**Draw a table on the board to record values for the variable *P* and the resulting number sentences.**	Using organizational tools

Vocabulary Development

The list below identifies the Key Vocabulary terms from this unit. The lesson in which each term is defined is indicated next to the term. Some of these terms or their homophones are used outside of mathematics. Consider adding other words as appropriate for developing understanding of the context of the lessons.

Lessons include suggestions for helping English language learners understand and develop vocabulary. For more information, see pages 17–19 of this handbook.

Key Vocabulary	
decimal point (*point) **4•5**	map scale (*scale) **4•3**
dividend **4•1, 4•2**	multiples **4•1**
†divisor **4•1, 4•2**	partial quotient **4•2**
magnitude estimate **4•5**	quotient **4•1**
map direction symbol (†symbol) **4•3**	remainder **4•2**
map legend, or map key (*legend, *†key) **4•3**	variable **4•7**

* Discuss the everyday and mathematical meanings of the words that are marked with an asterisk.

† For words marked with a dagger, write the words and their homophones on the board. For example, *divisor* and *deviser; symbol* and *cymbal;* and *key* and *quay*. Discuss and clarify the meaning of each.

◆ As each word is introduced in the lesson, write the word on the board and discuss its meaning.

◆ List the words on a Math Word Wall for students to see. As each word is introduced in the lesson, add a picture next to the word on the Word Wall.

◆ Use the vocabulary words regularly when teaching lessons, and encourage students to use the words in their discussions.

Games

Below are suggested Unit 4 game adaptations. For more information about implementing games in a differentiated classroom, see pages 20–25 of this handbook.

Game: *Division Dash*

Skill Practiced: Divide 2- or 3-digit dividends by 1-digit divisors. [Operations and Computation Goal 3]

Modification	Purpose of Modification
Players draw two cards to make a dividend. They choose a divisor that evenly divides their dividend. The score for a round is the divisor they choose. If the divisor does not evenly divide the dividend, there is no score for the round.	Students divide 2-digit dividends using a 1-digit divisor of their choice. [Operations and Computation Goal 3]
Players draw four cards. They make a 3-digit dividend and a 1-digit divisor greater than 2. They get ten bonus points if they make a problem with no remainder. Instead of playing to 100, they play five rounds, and the highest score wins.	Students divide 4-digit dividends by divisors greater than 2. [Operations and Computation Goal 3]

Game: *Division Top-It*

Skill Practiced: Find and compare quotients for division problems. [Number and Numeration Goal 6; Operations and Computation Goal 3]

Modification	Purpose of Modification
Players separate their cards into a divisor pile with four each of 2s, 5s, and 10s. They shuffle and reuse the divisor pile. The rest of the number cards go in the dividend pile. For each turn, players turn over one divisor card and two dividend cards. They make a 2-digit number with the dividend cards.	Students find and compare quotients for division problems with 2, 5, or 10 as the divisor. [Number and Numeration Goal 6; Operations and Computation Goal 3]
Players draw four cards and make a 3-digit dividend and 1-digit divisor. They receive a bonus point if they can form a division problem with no remainder.	Students find and compare quotients for division problems with 3-digit dividends. [Number and Numeration Goal 6; Operations and Computation Goal 3]

Game: *First to 100*

Skill Practiced: Solve open number sentences. [Patterns, Functions, and Algebra Goal 2]

Modification	Purpose of Modification
Players make their own set of *First to 100* cards. Each player makes four open number sentence cards similar to the *First to 100* cards—one each for addition, subtraction, multiplication, and division. Play ends after ten rounds. The player with the higher score wins.	Students solve open number sentences for the four basic operations. [Patterns, Functions, and Algebra Goal 2]
Players may multiply or divide the product of their dice by 10. Play ends after ten rounds. The score closest to 100 wins.	Students solve open number sentences involving multiples of 10. [Patterns, Functions, and Algebra Goal 2]

Math Boxes

Suggestions for using Math Boxes to meet individual needs begin on page 26 of this handbook. There are blank masters for Math Boxes on pages 136–141.

Using Part 3 of the Lessons

Use your professional judgment, along with assessment results, to determine whether the whole class, small groups, or individual students might benefit from these Unit 4 activities. Consider using the Part 3 Planning Master found on page 154 of this handbook to record your plans.

Readiness Activities

Lesson	Activity	Purpose of Activity
4•1	**Complete name-collection boxes.**	Gain experience finding equivalent names for numbers. [Number and Numeration Goal 4]
4•2	**Find factors of numbers using divisibility rules.**	Gain experience identifying factors. [Number and Numeration Goal 3]
4•3	**Review fraction measurements on a ruler and write rounded measures for line segments.**	Explore strategies for measuring and rounding to the nearest fraction of an inch. [Measurement and Reference Frames Goal 1]
4•4	**Write numbers in expanded notation.**	Explore using extended facts. [Number and Numeration Goal 1]
4•5	**Model and solve long-division problems with base-10 blocks.**	Gain experience with division. [Operations and Computation Goal 3]
4•6	**Identify and record the important information in number stories.**	Gain experience identifying the relevant information in number stories. [Operations and Computation Goal 2]
4•7	**Use situation diagrams and write open number sentences to solve number stories.**	Explore situation diagrams and their connection to number sentences. [Patterns, Functions, and Algebra Goal 2]

English Language Learners Support Activities

Lesson	Activity	Purpose of Activity
4•2	**Write a division model in several formats and label the *dividend, divisor, quotient,* and *remainder* for each.**	Use a visual model to clarify the mathematical uses of the terms. [Operations and Computation Goal 3]
4•5	**Make posters for different *division algorithms.***	Use student-made posters as visual references. [Operations and Computation Goal 3]
4•7	**Add *variable* to the Math Word Bank.**	Make connections among and use visuals to represent terms. [Patterns, Functions, and Algebra Goal 2]

Enrichment Activities

Lesson	Activity	Purpose of Activity
4•1	**Solve problems using divisibility rules for prime numbers.**	Apply understanding of divisibility. [Operations and Computation Goal 3]
4•2	**Identify 3-digit numbers that are divisible by specified numbers.**	Apply understanding of factors. [Number and Numeration Goal 3]
4•3	**Estimate distances of curved paths using a map scale.**	Apply understanding of estimating distances using a map scale. [Operations and Computation Goal 7]
4•5	**Use the column-division algorithm for decimal division.**	Explore division algorithms. [Operations and Computation Goal 3]
4•6	**Write and solve division number stories.**	Apply understanding of division. [Operations and Computation Goal 3]
4•7	**Play *Algebra Election*.**	Apply understanding of solving open number sentences. [Patterns, Functions, and Algebra Goal 2]

Extra Practice Activities

Lesson	Activity	Purpose of Activity
4•1	**Solve *5-Minute Math* problems involving division.**	Practice whole-number division. [Operations and Computation Goal 3]
4•3	**Identify and draw a route from New York to Los Angeles and measure the distance using a map scale.**	Practice estimating distances using a map scale. [Operations and Computation Goal 7]
4•4	**Solve division problems using lists of multiples of the divisor.**	Practice dividing. [Operations and Computation Goal 3]
4•6	**Solve *5-Minute Math* problems involving division.**	Practice solving division problems with remainders. [Operations and Computation Goal 3]
4•7	**Solve open number sentences.**	Practice solving open number sentences. [Patterns, Functions, and Algebra Goal 2]

Looking at Grade-Level Goals

Everyday Mathematics develops concepts and skills over time. Below is a chart showing where the Grade-Level Goals emphasized in this unit are addressed throughout the year. Use the chart to help you determine which Maintaining Concepts and Skills activities on page 77 to utilize to ensure that students continue working toward these Grade-Level Goals.

● Grade-Level Goal is taught.
◍ Grade-Level Goal is practiced and applied.
○ Grade-Level Goal is not a focus.

Grade-Level Goals Emphasized in Unit 4	Unit											
	1	2	3	4	5	6	7	8	9	10	11	12
[Operations and Computation Goal 2] Demonstrate automaticity with multiplication and division fact extensions.	●	◍	◍	●	○	○	○	○	○	○	○	○
[Operations and Computation Goal 3] Use manipulatives, mental arithmetic, paper-and-pencil algorithms and models, and calculators to solve problems involving the multiplication of whole numbers and decimals and the division of multidigit whole numbers and decimals by whole numbers; express remainders as whole numbers or fractions as appropriate; describe the strategies used and explain how they work.	●	●	◍	●	◍	◍	◍	◍	◍	◍	◍	◍
[Operations and Computation Goal 6] Make reasonable estimates for whole number and decimal addition, subtraction, multiplication, and division problems and fraction and mixed number addition and subtraction problems; explain how the estimates were obtained.	◍	●	◍	●	○	◍	○	◍	○	◍	○	○
[Measurement and Reference Frames Goal 1] Estimate length with and without tools; measure length with tools to the nearest $\frac{1}{8}$ inch and millimeter; estimate the measure of angles with and without tools; use tools to draw angles with given measures.	○	◍	●	●	●	●	◍	◍	◍	○	○	○
[Patterns, Functions, and Algebra Goal 2] Determine whether number sentences are true or false; solve open number sentences and explain the solutions; use a letter variable to write an open sentence to model a number story; use a pan-balance model to solve linear equations in one unknown.	○	◍	◍	●	◍	◍	●	◍	○	●	◍	◍

Maintaining Concepts and Skills

Some of the goals addressed in this unit will be addressed again in later units. Those goals marked with an asterisk (*) are addressed in future units only as practice and application. Here are several suggestions for maintaining concepts and skills until goals are revisited.

Operations and Computation Goal 2*

- Have students review divisibility rules for 1-digit divisors. See the Readiness activity in Lesson 4-2 for more information.
- Use the Name-Collection Boxes master on page 147 of this handbook to create practice problems. Have students record names using multiplication and division.

Operations and Computation Goal 3*

- Have students play *Division Dash* and *Division Top-It.*
- Have students model and solve division problems with base-10 blocks. See the Readiness activity in Lesson 4-5 for more information.

Operations and Computation Goal 6*

- Have students routinely make ballpark estimates for computation problems before solving the problems.

Patterns, Functions, and Algebra Goal 2

- Have students play *Algebra Election.*
- Have students solve for unknown quantities. See the Readiness activity in Lesson 4-7 for more information.

Assessment

See page 78 in the *Assessment Handbook* for modifications to the written portion of the Unit 4 Progress Check.

Additionally, see pages 79–83 for modifications to the open-response task and selected student work samples.

Unit 5

Activities and Ideas for Differentiation

In this unit, students describe, compare, and classify plane figures. This section summarizes opportunities for supporting multiple learning styles and ability levels. Use these suggestions to develop a differentiation plan for Unit 5.

Part 1 Activities That Support Differentiation

Below are examples of Unit 5 activities that highlight some of the general instructional strategies that are hallmarks of a differentiated classroom. These strategies will help you support, emphasize, and enhance lesson content to make sure all of your students are engaged in the mathematics at the highest possible level. For more information about general differentiation strategies that accommodate the diverse needs of today's classrooms, see the essay on pages 8–16 of this handbook.

Lesson	Activity	Strategy
5•1	**Students collect examples of fractions, decimals, and percents for a Fractions, Decimals, and Percents Museum.**	Connecting to everyday life
5•2	**Students use pattern blocks to model fraction and mixed-number problems.**	Modeling concretely
5•3	**Students use a Fraction-Stick Chart to compare fractions and to find equivalent fractions.**	Modeling visually
5•4	**Students divide rectangles to model equivalent fractions and formulate the multiplication rule for finding equivalent fractions.**	Modeling visually
5•7	**Students share their strategies for finding the decimal names for fractions.**	Incorporating and validating a variety of methods
5•9	**Students compare a bar graph and a circle graph.**	Talking about math

Vocabulary Development

The list below identifies the Key Vocabulary terms from this unit. The lesson in which each term is defined is indicated next to the term. Some of these terms or their homophones are used outside of mathematics. Consider adding other words as appropriate for developing understanding of the context of the lessons.

Lessons include suggestions for helping English language learners understand and develop vocabulary. For more information, see pages 17–19 of this handbook.

Key Vocabulary

bar graph (*bar) **5◆9**	percent **5◆8**
benchmark **5◆3**	Percent Circle **5◆10**
circle (or *†pie) graph **5◆9**	repeating decimal **5◆7**
denominator **5◆1**	round down (*round, *down) **5◆5**
fraction stick (*stick) **5◆3**	round to the nearest . . . **5◆5**
equivalent fractions **5◆3**	round up **5◆5**
improper fraction (*improper) **5◆2**	sector **5◆10**
mixed number (*mixed) **5◆2**	unit fraction (*unit) **5◆1**
numerator **5◆1**	†whole (†ONE, or unit) **5◆1**

* Discuss the everyday and mathematical meanings of the words that are marked with an asterisk.

† For words marked with a dagger, write the words and their homophones on the board. For example, *pie* and *pi; whole* and *hole;* and *one* and *won.* Discuss and clarify the meaning of each.

◆ As each word is introduced in the lesson, write the word on the board and discuss its meaning.

◆ List the words on a Math Word Wall for students to see. As each word is introduced in the lesson, add a picture next to the word on the Word Wall.

◆ Use the vocabulary words regularly when teaching lessons, and encourage students to use the words in their discussions.

Games

Below are suggested Unit 5 game adaptations. For more information about implementing games in a differentiated classroom, see pages 20–25 of this handbook.

Game: *Fraction Top-It*

Skill Practiced: Compare and order fractions. [Number and Numeration Goal 6]

Modification	Purpose of Modification
Players use only the cards showing halves, fourths, eighths, and twelfths.	Students compare and order fractions. [Number and Numeration Goal 6]
Players draw two fractions on each turn and compare the sums of their fraction pairs.	Students compare, order, and add fractions. [Number and Numeration Goal 6; Operations and Computation Goal 4]

Game: *Number Top-It*

Skill Practiced: Build and compare large numbers. [Number and Numeration Goals 1 and 6]

Modification	Purpose of Modification
Players place zeros in the ones, tens, and hundreds places. Then they draw four cards to fill the remaining spaces in the place-value chart.	Students build and compare numbers in the thousands or millions. [Number and Numeration Goals 1 and 6]
In each round, a player makes two numbers, and his or her score is the difference between the numbers. The highest score wins the round.	Students subtract large numbers and compare differences. [Number and Numeration Goal 6; Operations and Computation Goal 1]

Game: *Fraction/Percent Concentration*

Skill Practiced: Develop automaticity with "easy" fraction/percent equivalencies. [Number and Numeration Goal 5]

Modification	Purpose of Modification
Players use only tenths and the equivalent percents.	Students develop automaticity with fraction/percent equivalencies for tenths. [Number and Numeration Goal 5]
Players receive a bonus point if they can name the equivalent decimal for their fraction/percent pairs. At the end of the game, bonus points count as additional cards.	Students develop automaticity with "easy" fraction/decimal/percent equivalencies. [Number and Numeration Goal 5]

Math Boxes

Suggestions for using Math Boxes to meet individual needs begin on page 26 of this handbook. There are blank masters for Math Boxes on pages 136–141.

Using Part 3 of the Lessons

Use your professional judgment, along with assessment results, to determine whether the whole class, small groups, or individual students might benefit from these Unit 5 activities. Consider using the Part 3 Planning Master found on page 154 of this handbook to record your plans.

Readiness Activities

Lesson	Activity	Purpose of Activity
5•1	**Identify missing numbers in number stories.**	Explore relationships between fractions of a set and fractions of a whole. [Number and Numeration Goal 2]
5•2	**Use pattern blocks to model relationships between fractions.**	Explore relationships between fractions. [Number and Numeration Goal 2]
5•3	**Make fraction strips and use them to compare fractions.**	Explore comparing and ordering fractions. [Number and Numeration Goal 6]
5•4	**Use the Everything Math Deck to identify equivalent fractions.**	Explore equivalent fractions. [Number and Numeration Goal 5]
5•5	**Round whole numbers and decimals using a number line.**	Explore rounding numbers. [Operations and Computation Goal 6]
5•6	**Play *Number Top-It.***	Gain experience with decimal place value and ordering decimals. [Number and Numeration Goals 1 and 6]
5•7	**Convert dictated decimal numbers to fractions.**	Gain experience with decimal place value. [Number and Numeration Goals 1 and 5]
5•8	**Play *Fraction/Percent Concentration.***	Gain experience converting between fractions and percents. [Number and Numeration Goal 5]
5•9	**Make a human circle graph.**	Gain experience with circle graphs. [Data and Chance Goal 1]
5•10	**Make fraction and percent references for circle-graph sectors.**	Gain experience with fraction and percent names for circle sectors. [Number and Numeration Goal 5]
5•11	**Use a Percent Circle to measure sectors in a circle graph.**	Explore measuring sectors of a circle graph. [Data and Chance Goal 1]

English Language Learners Support Activities

Lesson	Activity	Purpose of Activity
5•1	**Describe how *fractions, decimals,* and *percents* are used in the Fractions, Decimals, and Percents Museum.**	Make connections between mathematics and everyday life; discuss new mathematical ideas. [Number and Numeration Goal 5]
5•3	**Add *equivalent fractions* to the Math Word Bank.**	Make connections among and use visuals to represent terms. [Number and Numeration Goal 5]
5•8	**Add *percent* to the Math Word Bank.**	Make connections among and use visuals to represent terms. [Number and Numeration Goal 5]
5•9	**Use a Venn diagram to compare *bar graphs* and *circle graphs.***	Use a graphic organizer to understand the terms. [Data and Chance Goal 1]

Enrichment Activities

Lesson	Activity	Purpose of Activity
5◆1	**Identify missing numbers in number stories.**	Explore fraction, whole-number, and mixed-number relationships. [Number and Numeration Goal 2]
5◆2	**Calculate fractional values of pattern-block shapes.**	Apply understanding of fractions. [Number and Numeration Goal 2]
5◆3	**Use a Fraction-Stick Chart to explore the relationships between numerators and denominators of equivalent fractions.**	Apply understanding of equivalent fractions. [Number and Numeration Goal 5]
5◆4	**Use the division rule to write fractions in simplest form.**	Apply understanding of equivalent fractions. [Number and Numeration Goal 5]
5◆6	**Write fraction and decimal equivalents for shaded 100-grids.**	Apply understanding of fractions and decimals. [Number and Numeration Goal 5]
5◆7	**Use the partial-quotients division algorithm to find decimal equivalents for fractions.**	Explore converting between fractions and decimals. [Number and Numeration Goal 5]
5◆8	**Write and solve "percent-of" number stories.**	Apply understanding of percents. [Number and Numeration Goal 2]
5◆10	**Conduct an eye-test experiment.**	Apply understanding of fractions, decimals, and percents. [Number and Numeration Goal 5]
5◆11	**Calculate percents from experiment data.**	Apply understanding of percents. [Number and Numeration Goal 5]
5◆12	**Read about the history of mathematics instruction and solve related problems.**	Explore proportional reasoning. [Operations and Computation Goal 7]

Extra Practice Activities

Lesson	Activity	Purpose of Activity
5◆2	**Use counters to solve "fraction-of" problems.**	Practice solving "fraction-of" problems. [Number and Numeration Goal 5]
5◆4	**Solve *5-Minute Math* problems involving equivalent fractions.**	Practice finding equivalent fractions. [Number and Numeration Goal 5]
5◆5	**Rename fractions and mixed numbers as decimals.**	Practice renaming fractions and mixed numbers as decimals. [Number and Numeration Goal 5]
5◆6	**Solve *5-Minute Math* problems involving fractions, decimals, and percents.**	Practice comparing fractions, decimals, and percents. [Number and Numeration Goal 6]
5◆8	**Write and solve number stories involving whole and mixed numbers, fractions, decimals, and percents.**	Practice solving number stories. [Number and Numeration Goal 2]
5◆9	**Find equivalent fractions using the multiplication and division rules.**	Practice finding equivalent fractions. [Number and Numeration Goal 5]
5◆10	**Convert between fractions and decimals.**	Practice converting between fractions and decimals. [Number and Numeration Goal 5]
5◆12	**Convert bar graphs to circle graphs.**	Practice constructing circle graphs. [Data and Chance Goal 1]

Looking at Grade-Level Goals

Everyday Mathematics develops concepts and skills over time. Below is a chart showing where the Grade-Level Goals emphasized in this unit are addressed throughout the year. Use the chart to help you determine which Maintaining Concepts and Skills activities on page 84 to utilize to ensure that students continue working toward these Grade-Level Goals.

- ● Grade-Level Goal is taught.
- ◍ Grade-Level Goal is practiced and applied.
- ○ Grade-Level Goal is not a focus.

Grade-Level Goals Emphasized in Unit 5	Unit											
	1	2	3	4	5	6	7	8	9	10	11	12
[Number and Numeration Goal 1] Read and write whole numbers and decimals; identify places in such numbers and the values of the digits in those places; use expanded notation to represent whole numbers and decimals.	◍	●	◍	◍	◍	◍	●	◍	◍	◍	○	○
[Number and Numeration Goal 2] Solve problems involving percents and discounts; describe and explain strategies used; identify the unit whole in situations involving fractions.	○	○	◍	○	●	◍	◍	●	○	○	◍	◍
[Number and Numeration Goal 5] Use numerical expressions to find and represent equivalent names for fractions, decimals, and percents; use and explain multiplication and division rules to find equivalent fractions and fractions in simplest form; convert between fractions and mixed numbers; convert between fractions, decimals, and percents.	◍	◍	○	◍	●	●	◍	●	◍	○	◍	◍
[Number and Numeration Goal 6] Compare and order rational numbers; use area models, benchmark fractions, and analyses of numerators and denominators to compare and order fractions and mixed numbers; describe strategies used to compare fractions and mixed numbers.	◍	◍	◍	◍	●	◍	●	●	◍	○	○	◍
[Operations and Computation Goal 4] Use mental arithmetic, paper-and-pencil algorithms and models, and calculators to solve problems involving the addition and subtraction of fractions and mixed numbers; describe the strategies used and explain how they work.	○	○	○	○	●	●	◍	●	◍	◍	○	○
[Data and Chance Goal 1] Collect and organize data or use given data to create graphic displays with reasonable titles, labels, keys, and intervals.	○	◍	○	○	●	●	◍	◍	○	◍	◍	○
[Measurement and Reference Frames Goal 1] Estimate length with and without tools; measure length with tools to the nearest $\frac{1}{8}$ inch and millimeter; estimate the measure of angles with and without tools; use tools to draw angles with given measures.	○	◍	●	●	●	●	◍	◍	◍	○	○	○

Maintaining Concepts and Skills

All of the goals addressed in this unit will be addressed again in later units. Here are several suggestions for maintaining concepts and skills until goals are revisited.

Number and Numeration Goal 2

- Have students play *Fraction Of.*
- Have students find fractions of a whole with pattern blocks. See the Readiness activity in Lesson 5-2 for more information.
- Use the "What's My Rule?" master on page 146 of this handbook to create practice problems in which the rule is to find a fraction of a number.

Number and Numeration Goal 5

- Have students play *Frac-Tac-Toe* and *Fraction/Percent Concentration.*
- Have students explore equivalent fractions. See the Readiness activity in Lesson 5-4 for more information.
- Have students routinely record the probability of events on the Probability Meter. The events can be listed on the Class Data Pad along with their probability expressed in words and as a fraction, decimal, and percent.

Number and Numeration Goal 6

- Have students play *Fraction Top-It* and *Number Top-It* (3-Place Decimals).

Measurement and Reference Frames Goal 1

- Have students play *Angle Tangle.*
- Have students measure sectors in a circle graph using a Percent Circle. See the Readiness activity in Lesson 5-11 for more information.

Assessment

See page 86 in the *Assessment Handbook* for modifications to the written portion of the Unit 5 Progress Check.

Additionally, see pages 87–91 for modifications to the open-response task and selected student work samples.

Unit 6 Activities and Ideas for Differentiation

In this unit, students extend their work with data and investigate the effect of sample size. They also review fraction concepts including adding and subtracting fractions, and finding equivalent fractions. This section summarizes opportunities for supporting multiple learning styles and ability levels. Use these suggestions to develop a differentiation plan for Unit 6.

Part 1 Activities That Support Differentiation

Below are examples of Unit 6 activities that highlight some of the general instructional strategies that are hallmarks of a differentiated classroom. These strategies will help you support, emphasize, and enhance lesson content to make sure all of your students are engaged in the mathematics at the highest possible level. For more information about general differentiation strategies that accommodate the diverse needs of today's classrooms, see the essay on pages 8–16 of this handbook.

Lesson	Activity	Strategy
6◆1	**Students make a stick-on-note line plot for number of states visited, and they identify data landmarks.**	Modeling concretely
6◆2	**Students measure the lengths of their *natural measures* using standard units.**	Connecting to everyday life
6◆3	**Students organize data for lengths of hand spans into stem-and-leaf plots.**	Using organizational tools
6◆5	**Students make circle graphs for colors of candy samples.**	Modeling concretely
6◆9	**Students use fractions on a clock face to add and subtract fractions.**	Building on prior knowledge
6◆10	**Students split fraction sticks to find common denominators.**	Modeling visually

Vocabulary Development

The list below identifies the Key Vocabulary terms from this unit. The lesson in which each term is defined is indicated next to the term. Some of these terms or their homophones are used outside of mathematics. Consider adding other words as appropriate for developing understanding of the context of the lessons.

Lessons include suggestions for helping English language learners understand and develop vocabulary. For more information, see pages 17–19 of this handbook.

Key Vocabulary

angle of separation **6◆3**
climate **6◆7**
common denominator (*common) **6◆9**
contour line **6◆7**
contour map **6◆7**
cubit **6◆2**
fair game (*†fair) **6◆2**
*fathom **6◆2**
frequency table (*table) **6◆6**
great span (*†great) **6◆3**
*landmark **6◆1**
*leaf **6◆3**
line plot (*plot) **6◆1**
map legend (map key) (*legend, *†key) **6◆7**
maximum **6◆1**
*median **6◆1**
minimum **6◆1**
*†mode **6◆1**
normal span (*normal) **6◆3**
population **6◆5**
precipitation **6◆7**
quick common denominator **6◆10**
sample **6◆5**
simplest form (*form) **6◆10**
span **6◆3**
*stem **6◆3**
stem-and-leaf plot **6◆3**
survey **6◆6**
unlike denominators **6◆9**

* Discuss the everyday and mathematical meanings of the words that are marked with an asterisk.

† For words marked with a dagger, write the words and their homophones on the board. For example, *fair* and *fare; great* and *grate; key* and *quay;* and *mode* and *mowed.* Discuss and clarify the meaning of each.

- As each word is introduced in the lesson, write the word on the board and discuss its meaning.
- List the words on a Math Word Wall for students to see. As each word is introduced in the lesson, add a picture next to the word on the Word Wall.
- Use the vocabulary words regularly when teaching lessons, and encourage students to use the words in their discussions.

Games

Below are suggested Unit 6 game adaptations. For more information about implementing games in a differentiated classroom, see pages 20–25 of this handbook.

Game: *Frac-Tac-Toe*

Skill Practiced: Find equivalent names for fractions and decimals. [Number and Numeration Goal 5]

Modification	Purpose of Modification
Players use a 2-4-5-10 gameboard (decimal or percent) and use only 10s in the denominator pile.	Students find equivalent names for tenths. [Number and Numeration Goal 5]
Players add one each of the numbers 12 and 20 to the denominator pile. Change the middle space of the gameboard to "Name the Decimal." Players can cover this space if they have a 12 or 20 in the denominator and correctly name the decimal equivalent for the fraction.	Students find equivalent names for fractions and decimals. [Number and Numeration Goal 5]

Game: *Divisibility Dash*

Skill Practiced: Divide 2- or 3-digit dividends by 1-digit divisors. [Number and Numeration Goal 3]

Modification	Purpose of Modification
Players include only the 2 and 5 cards in the divisor pile.	Students divide 2-digit dividends using only 2 and 5 as divisors. [Number and Numeration Goal 3]
Each divisor card is multiplied by 10. If a 3 is turned over in the divisor pile, 30 is the divisor. Players have a permanent 0 card (that cannot be discarded) and make a 3- or 4-digit dividend for each divisor using the 0 card. For example, a player draws 3, 2, and 4 cards. The player makes the number 3,240 as the dividend that is a multiple of 30.	Students divide 3- or 4-digit dividends by divisors that are multiples of 10. [Number and Numeration Goal 3]

Game: *Fraction Capture*

Skill Practiced: Find equivalent names for fractions. [Number and Numeration Goal 5; Operations and Computation Goal 4]

Modification	Purpose of Modification
Players use the fraction side of the Everything Math Deck cards for halves, thirds, fourths, fifths, sixths, eighths, tenths, and twelfths to determine their fractions. If they draw a fraction card that cannot be covered on the board, they lose their turn.	Students find equivalent names for fractions. [Number and Numeration Goal 5; Operations and Computation Goal 4]
Players must capture *all* of the fraction pieces to end the game. Players need to adjust their strategies to use some of their moves to cover fraction pieces that will not win them squares.	Students find equivalent names for fractions. [Number and Numeration Goal 5; Operations and Computation Goal 4]

Math Boxes

Suggestions for using Math Boxes to meet individual needs begin on page 26 of this handbook. There are blank masters for Math Boxes on pages 136–141.

Using Part 3 of the Lessons

Use your professional judgment, along with assessment results, to determine whether the whole class, small groups, or individual students might benefit from these Unit 6 activities. Consider using the Part 3 Planning Master found on page 154 of this handbook to record your plans.

Readiness Activities

Lesson	Activity	Purpose of Activity
6•1	**Identify data landmarks in snack-survey data.**	Gain experience with definitions of and methods for finding data landmarks. [Data and Chance Goal 2]
6•2	**Measure and draw line segments using centimeters and millimeters.**	Explore the relationship between centimeters and millimeters. [Measurement and Reference Frames Goals 1 and 3]
6•3	**Measure angles with a half-circle protractor.**	Gain experience using a half-circle protractor. [Measurement and Reference Frames Goal 1]
6•4	**Use grid-paper strips to find the median of a data set.**	Explore finding medians. [Data and Chance Goal 2]
6•5	**Identify parts and wholes in number stories and convert the fractions to percents.**	Explore converting between fractions and percents. [Number and Numeration Goal 5]
6•6	**List data sets from four stem-and-leaf plots.**	Gain experience identifying individual values in stem-and-leaf plots. [Data and Chance Goal 1]
6•8	**Sort fraction cards using benchmarks and number sense.**	Compare fractions to $\frac{1}{2}$. [Number and Numeration Goal 6]

English Language Learners Support Activities

Lesson	Activity	Purpose of Activity
6•1	**Create a line-plot poster and discuss the landmarks *minimum, maximum, range, mode, median, mean,* and the term *outlier.***	Use student-made posters as a visual reference for new terms. [Data and Chance Goal 2]
6•3	**Add *stem-and-leaf plot* to the Math Word Bank.**	Make connections among and use visuals to represent terms. [Data and Chance Goal 1]
6•6	**Add *sample* and *population* to the Math Word Bank.**	Make connections among and use visuals to represent terms. [Data and Chance Goal 4]
6•7	**Make a graphic organizer for *climate.***	Connect a new term to existing vocabulary; use a graphic organizer to describe the characteristics of *climate.* [Measurement and Reference Frames Goal 3]

Enrichment Activities

Lesson	Activity	Purpose of Activity
6◆1	Create a line plot with fractional units.	Apply understanding of line plots in displaying and interpreting data. [Data and Chance Goals 1 and 2]
6◆2	Predict the relationship between palm widths and joint lengths.	Apply understanding of data landmarks. [Data and Chance Goal 2]
6◆4	Analyze spelling-test scores.	Apply understanding of the relationship between median and mean. [Data and Chance Goal 2]
6◆5	Conduct an experiment with random outcomes and investigate how sample size affects results.	Apply understanding of the relationship between sample size and reliability of predictions. [Data and Chance Goal 2]
6◆7	Write and answer questions involving contour maps.	Explore analyzing contour maps. [Data and Chance Goal 2]
6◆8	Play *Fraction Top-It* (Advanced Version).	Apply knowledge of benchmarks. [Number and Numeration Goal 6]
6◆9	Model fractions and fraction operations with a 24-hour military clock.	Apply understanding of fractions on a clock face. [Operations and Computation Goal 4]
6◆10	Use the least-common-multiple method to find common denominators.	Apply understanding of common denominators. [Number and Numeration Goal 5]

Extra Practice Activities

Lesson	Activity	Purpose of Activity
6◆1	Solve *5-Minute Math* problems involving medians.	Practice finding the median for a set of data. [Data and Chance Goal 2]
6◆3	Organize collected data into a stem-and-leaf plot.	Practice organizing data. [Data and Chance Goal 1]
6◆4	Match line plots to data descriptions.	Practice relating descriptions of data to representations. [Data and Chance Goal 2]
6◆6	Make stem-and-leaf plots for provided data and find data landmarks.	Practice making stem-and-leaf plots and finding data landmarks. [Data and Chance Goals 1 and 2]
6◆7	Examine and discuss contour maps.	Practice with contour maps. [Data and Chance Goal 2]
6◆8	Write fraction number stories.	Practice adding and subtracting fractions. [Operations and Computation Goal 4]
6◆9	Write and solve elapsed-time number stories using fractions.	Practice solving fraction number stories. [Operations and Computation Goal 4]
6◆10	Find common denominators to add fractions.	Practice adding fractions. [Operations and Computation Goal 4]

Looking at Grade-Level Goals

Everyday Mathematics develops concepts and skills over time. Below is a chart showing where the Grade-Level Goals emphasized in this unit are addressed throughout the year. Use the chart to help you determine which Maintaining Concepts and Skills activities on page 91 to utilize to ensure that students continue working toward these Grade-Level Goals.

● Grade-Level Goal is taught.
◍ Grade-Level Goal is practiced and applied.
○ Grade-Level Goal is not a focus.

Grade-Level Goals Emphasized in Unit 6	Unit											
	1	2	3	4	5	6	7	8	9	10	11	12
[Operations and Computation Goal 4] Use mental arithmetic, paper-and-pencil algorithms and models, and calculators to solve problems involving the addition and subtraction of fractions and mixed numbers; describe the strategies used and explain how they work.	○	○	○	○	●	●	◍	●	◍	◍	○	○
[Data and Chance Goal 1] Collect and organize data or use given data to create graphic displays with reasonable titles, labels, keys, and intervals.	○	◍	○	○	●	●	◍	◍	○	◍	◍	○
[Data and Chance Goal 2] Use the maximum, minimum, range, median, mode, and mean and graphs to ask and answer questions, draw conclusions, and make predictions.	◍	●	●	◍	◍	●	◍	○	◍	◍	○	○
[Data and Chance Goal 4] Predict the outcomes of experiments, test the predictions using manipulatives, and summarize the results; compare predictions based on theoretical probability with experimental results; use summaries and comparisons to predict future events; express the probability of an event as a fraction, decimal, or percent.	◍	●	○	○	◍	●	◍	○	◍	○	○	◍
[Measurement and Reference Frames Goal 1] Estimate length with and without tools; measure length with tools to the nearest $\frac{1}{8}$ inch and millimeter; estimate the measure of angles with and without tools; use tools to draw angles with given measures.	○	◍	●	●	●	●	◍	◍	◍	○	○	○

Maintaining Concepts and Skills

Some of the goals addressed in this unit will be addressed again in later units. Those goals marked with an asterisk (*) are addressed in future units only as practice and application. Here are several suggestions for maintaining concepts and skills until goals are revisited.

Number and Numeration Goal 3*

- Have students play *Factor Captor* and *Factor Bingo.*
- Use the Name-Collection Boxes master on page 147 of this handbook to create practice problems. Have students record prime-factorization names for numbers.

Operations and Computation Goal 4

- Have students play *Fraction Capture.* Have them record number sentences when they shade areas in more than one square on a turn.
- Use the "What's My Rule?" master on page 146 of this handbook to create practice problems in which the rule is to add or subtract a fraction or mixed number.

Data and Chance Goal 2*

- Have students review data landmarks using survey data. See the Readiness activity in Lesson 6-1 for more information.
- Have students explore stem-and-leaf data. See the Readiness activity in Lesson 6-6 for more information.

Measurement and Reference Frames Goal 1*

- Have students play *Angle Tangle.*
- Have students measure and draw line segments. See the Readiness activity in Lesson 6-2 for more information.
- Have students measure angles. See the Readiness activity in Lesson 6-3 for more information.

Assessment

See page 94 in the *Assessment Handbook* for modifications to the written portion of the Unit 6 Progress Check.

Additionally, see pages 95–99 for modifications to the open-response task and selected student work samples.

Unit 7 Activities and Ideas for Differentiation

In this unit, students further develop their pre-algebra skills, use of parentheses, and work with negative numbers. This section summarizes opportunities for supporting multiple learning styles and ability levels. Use these suggestions to develop a differentiation plan for Unit 7.

Part 1 Activities That Support Differentiation

Below are examples of Unit 7 activities that highlight some of the general instructional strategies that are hallmarks of a differentiated classroom. These strategies will help you support, emphasize, and enhance lesson content to make sure all of your students are engaged in the mathematics at the highest possible level. For more information about general differentiation strategies that accommodate the diverse needs of today's classrooms, see the essay on pages 8–16 of this handbook.

Lesson	Activity	Strategy
7•1	**Students make true statements about standard notation and exponential notation.**	Talking about math
7•2	**Students use a place-value chart to compare powers of 10.**	Using organizational tools
7•4	**Record the steps used to evaluate numeric expressions on the board.**	Recording key ideas
7•5	**Record the rules for order of operations on the board.**	Using a visual reference
7•7	**Students discuss real-world examples of negative numbers.**	Connecting to everyday life

Vocabulary Development

The list below identifies the Key Vocabulary terms from this unit. The lesson in which each term is defined is indicated next to the term. Some of these terms or their homophones are used outside of mathematics. Consider adding other words as appropriate for developing understanding of the context of the lessons.

Lessons include suggestions for helping English language learners understand and develop vocabulary. For more information, see pages 17–19 of this handbook.

Key Vocabulary

account balance (*account, *balance) **7◆8**
ambiguous **7◆4**
axis **7◆6**
*†base **7◆1**
change-sign key (*change, *sign, *†key) **7◆11**
debt **7◆8**
expanded notation **7◆3**
exponent **7◆1**
exponential notation **7◆1**
*expression **7◆4**
*factor **7◆1**
in the black (*black) **7◆8**
in the red (*†red) **7◆8**
line graph **7◆6**
negative number (*negative) **7◆7**
nested parentheses (*nested) **7◆4**
number-and-word notation **7◆2**
opposite **7◆11**
order of operations (*order) **7◆5**
power of a number (*power) **7◆1**
powers of 10 **7◆2**
scientific notation **7◆3**
standard notation **7◆1**
trend **7◆6**
Venn diagram **7◆6**

* Discuss the everyday and mathematical meanings of the words that are marked with an asterisk.

† For words marked with a dagger, write the words and their homophones on the board. For example, *base* and *bass; key* and *quay;* and *red* and *read.* Discuss and clarify the meaning of each.

◆ As each word is introduced in the lesson, write the word on the board and discuss its meaning.

◆ List the words on a Math Word Wall for students to see. As each word is introduced in the lesson, add a picture next to the word on the Word Wall.

◆ Use the vocabulary words regularly when teaching lessons, and encourage students to use the words in their discussions.

Games

Below are suggested Unit 7 game adaptations. For more information about implementing games in a differentiated classroom, see pages 20–25 of this handbook.

Game: *Name That Number*

Skill Practiced: **Find equivalent names for numbers.** [Number and Numeration Goal 4; Patterns, Functions, and Algebra Goal 3]

Modification	Purpose of Modification
Players keep the same target number for each round; for example, 10 is always the target number.	Students find equivalent names for one target number. [Number and Numeration Goal 4]
The card turned over for the target is multiplied by 10 to find the target number. For example, if an 8 is turned over, the target is 80.	Students find equivalent names for multiples of 10. [Number and Numeration Goal 4]

Game: *Credits/Debits Game*

Skill Practiced: **Add positive and negative numbers.** [Operations and Computation Goal 1]

Modification	Purpose of Modification
Players use a number line to model the addition and subtraction problems.	Students add positive and negative numbers on a number line. [Operations and Computation Goal 1]
Players record an addition number sentence for each round of the game.	Students add positive and negative numbers, and record number sentences. [Operations and Computation Goal 1; Patterns, Functions, and Algebra Goal 2]

Game: *High-Number Toss: Decimal Version*

Skill Practiced: **Identify place value and compare decimals.** [Number and Numeration Goals 1 and 6]

Modification	Purpose of Modification
For each round, players draw only 2 cards from the deck to create a decimal to the hundredths place.	Students read, write, and compare decimals. [Number and Numeration Goals 1 and 6]
Students receive a bonus point if they can round their decimal number to the nearest tenth.	Students read, write, compare, and round decimals. [Number and Numeration Goals 1 and 6; Operations and Computation Goal 6]

Math Boxes

Suggestions for using Math Boxes to meet individual needs begin on page 26 of this handbook. There are blank masters for Math Boxes on pages 136–141.

Using Part 3 of the Lessons

Use your professional judgment, along with assessment results, to determine whether the whole class, small groups, or individual students might benefit from these Unit 7 activities. Consider using the Part 3 Planning Master found on page 154 of this handbook to record your plans.

Readiness Activities

Lesson	Activity	Purpose of Activity
7•1	**Find patterns in numbers written in exponential notation.**	Explore the meaning of exponents. [Number and Numeration Goal 1]
7•2	**Find patterns in a place-value chart.**	Explore place value. [Number and Numeration Goal 4; Patterns, Functions, and Algebra Goal 1]
7•3	**Complete name-collection boxes.**	Explore the use of place value and number-and-word notation to rename numbers. [Number and Numeration Goals 1 and 4]
7•4	**Insert parentheses to make number sentences true.**	Explore the use of parentheses in number sentences. [Patterns, Functions, and Algebra Goal 3]
7•5	**Compare how 4-function and scientific calculators evaluate expressions.**	Explore rules for order of operations. [Patterns, Functions, and Algebra Goal 3]
7•6	**Analyze line graphs from newspapers or magazines.**	Gain experience with line graphs. [Data and Chance Goal 2]
7•7	**Skip count back on a calculator to explore negative numbers.**	Gain experience with negative numbers. [Number and Numeration Goal 1]
7•8	**Play the *Credits/Debits Game.***	Explore adding signed numbers. [Operations and Computation Goal 1]
7•10	**Label fractions on number lines.**	Gain experience with fractional units on a number line. [Number and Numeration Goal 6]

English Language Learners Support Activities

Lesson	Activity	Purpose of Activity
7•1	**Add *prime factorization, base, exponent,* and *power* to the Math Word Bank.**	Make connections among and use visuals to represent terms. [Number and Numeration Goals 1 and 3]
7•3	**Use a Venn diagram to compare and contrast *standard notation* and *exponential notation.***	Use a graphic organizer to clarify the mathematical uses of the terms. [Number and Numeration Goal 1]
7•8	**Make a visual aid for the accounting terms *account balance, in the black, in the red, debt,* and *cash.***	Use a teacher-made poster as a visual reference for terms. [Operations and Computation Goal 1]
7•11	**Add *negative numbers* to the Math Word Bank.**	Make connections among and use visuals to represent terms. [Operations and Computation Goal 1]

Enrichment Activities

Lesson	Activity	Purpose of Activity
7•1	**Use patterns to solve problems involving numbers with exponents and Fibonacci numbers.**	Apply understanding of exponents. [Patterns, Functions, and Algebra Goal 1]
7•2	**Use patterns to convert between numbers with negative exponents and standard notation.**	Apply understanding of exponents. [Number and Numeration Goal 4]
7•4	**Write number sentences to describe dot patterns.**	Apply understanding of parentheses. [Patterns, Functions, and Algebra Goal 3]
7•5	**Solve problems that involve exponents and operations.**	Apply understanding of order of operations. [Patterns, Functions, and Algebra Goal 3]
7•7	**Calculate price changes using positive and negative fractions and decimals.**	Apply understanding of negative numbers. [Number and Numeration Goal 5; Operations and Computation Goal 1]
7•8	**Play a game of *500*.**	Apply understanding of addition of negative numbers. [Operations and Computation Goal 1]
7•9	**Compare U.S. locations that are above and below sea level.**	Apply understanding of adding and subtracting signed numbers. [Operations and Computation Goal 1]
7•11	**Play *Broken Calculator,* using addition and subtraction of negative numbers.**	Apply understanding of operations with signed numbers. [Operations and Computation Goal 1]

Extra Practice Activities

Lesson	Activity	Purpose of Activity
7•1	**Calculate and express the number of possible 4-character computer passwords using exponential notation.**	Practice using exponential notation. [Number and Numeration Goal 4]
7•2	**Solve problems involving the multiplication of decimals by powers of 10.**	Practice multiplying decimals by powers of 10. [Operations and Computation Goal 3]
7•3	**Use addition and multiplication expressions to write whole numbers and decimals in expanded notation.**	Practice writing whole numbers and decimals in expanded notation. [Number and Numeration Goal 1]
7•4	**Solve *5-Minute Math* problems involving grouping symbols.**	Practice using grouping symbols. [Patterns, Functions, and Algebra Goal 3]
7•6	**Graph sets of temperature data on a line graph.**	Practice making line graphs for data. [Data and Chance Goal 1]
7•7	**Solve *5-Minute Math* problems involving negative numbers.**	Practice solving problems with negative numbers. [Operations and Computation Goal 1]
7•9	**Solve *5-Minute Math* "What's My Rule?" problems involving negative numbers.**	Practice solving problems with negative numbers. [Patterns, Functions, and Algebra Goal 1]
7•10	**Use rain gauge data to create a line plot.**	Practice creating and analyzing a line plot with fractional units. [Data and Chance Goals 1 and 2]

Looking at Grade-Level Goals

Everyday Mathematics develops concepts and skills over time. Below is a chart showing where the Grade-Level Goals emphasized in this unit are addressed throughout the year. Use the chart to help you determine which Maintaining Concepts and Skills activities on page 98 to utilize to ensure that students continue working toward these Grade-Level Goals.

- ● Grade-Level Goal is taught.
- ◍ Grade-Level Goal is practiced and applied.
- ○ Grade-Level Goal is not a focus.

Grade-Level Goals Emphasized in Unit 7	Unit											
	1	2	3	4	5	6	7	8	9	10	11	12
[Number and Numeration Goal 1] Read and write whole numbers and decimals; identify places in such numbers and the values of the digits in those places; use expanded notation to represent whole numbers and decimals.	◍	●	◍	◍	◍	◍	●	◍	◍	◍	○	○
[Number and Numeration Goal 4] Use numerical expressions involving one or more of the basic four arithmetic operations, grouping symbols, and exponents to give equivalent names for whole numbers; convert between base-10, exponential, and repeated-factor notations.	●	◍	◍	○	○	○	●	○	○	◍	◍	○
[Number and Numeration Goal 6] Compare and order rational numbers; use area models, benchmark fractions, and analyses of numerators and denominators to compare and order fractions and mixed numbers; describe strategies used to compare fractions and mixed numbers.	◍	◍	◍	◍	●	◍	●	●	◍	○	○	◍
[Operations and Computation Goal 1] Use manipulatives, mental arithmetic, paper-and-pencil algorithms and models, and calculators to solve problems involving the addition and subtraction of whole numbers, decimals, and signed numbers; describe the strategies used and explain how they work.	◍	●	◍	◍	◍	◍	●	◍	◍	◍	◍	○
[Patterns, Functions, and Algebra Goal 1] Extend, describe, and create numeric patterns; describe rules for patterns and use them to solve problems; write rules for functions involving the four basic arithmetic operations; represent functions using words, symbols, tables, and graphs and use those representations to solve problems.	●	○	◍	○	○	○	●	●	◍	●	○	◍
[Patterns, Functions, and Algebra Goal 2] Determine whether number sentences are true or false; solve open number sentences and explain the solutions; use a letter variable to write an open sentence to model a number story; use a pan-balance model to solve linear equations in one unknown.	○	◍	◍	●	◍	◍	●	◍	○	●	◍	◍
[Patterns, Functions, and Algebra Goal 3] Evaluate numeric expressions containing grouping symbols and nested grouping symbols; insert grouping symbols and nested grouping symbols to make number sentences true; describe and use the precedence of multiplication and division over addition and subtraction.	○	○	○	◍	○	◍	●	◍	○	○	○	◍

Maintaining Concepts and Skills

Those goals marked with an asterisk (*) are addressed in future units only as practice and application. Here are several suggestions for maintaining concepts and skills until goals are revisited.

Number and Numeration Goal 1*

- Have students play *High-Number Toss.*
- Have students find patterns in the place-value chart. See the Readiness activity in Lesson 7-2 for more information.

Number and Numeration Goal 4*

- Have students play *Name That Number.*
- Use the Name-Collection Boxes master on page 147 of this handbook to create practice problems for whole numbers.

Operations and Computation Goal 1*

- Have students play *500* and *Broken Calculator.*

Patterns, Functions, and Algebra Goal 3*

- Have students play *Name That Number* and record number sentences for their solutions that correctly employ order of operations.
- Have students explore the need for order of operations. See the Readiness activity in Lesson 7-5 for more information.

Assessment

See page 104 in the *Assessment Handbook* for modifications to the written portion of the Unit 7 Progress Check.

Additionally, see pages 105–109 for modifications to the open-response task and selected student work samples.

Unit 8 Activities and Ideas for Differentiation

In this unit, students review finding equivalent fractions and further develop their computation skills with fractions and percents. This section summarizes opportunities for supporting multiple learning styles and ability levels. Use these suggestions to develop a differentiation plan for Unit 8.

Part 1 Activities That Support Differentiation

Below are examples of Unit 8 activities that highlight some of the general instructional strategies that are hallmarks of a differentiated classroom. These strategies will help you support, emphasize, and enhance lesson content to make sure all of your students are engaged in the mathematics at the highest possible level. For more information about general differentiation strategies that accommodate the diverse needs of today's classrooms, see the essay on pages 8–16 of this handbook.

Lesson	Activity	Strategy
8♦1	**Students use the Fraction-Stick and Decimal Number-Line Chart to find equivalent fractions.**	Modeling visually
8♦3	**Students compare solving fraction subtraction problems that require renaming of the fraction parts with problems that do not require renaming.**	Talking about math
8♦4	**Students use fraction cards to order fractions.**	Modeling visually
8♦6	**Students fold paper to create an area model for multiplying fractions.**	Modeling physically
8♦8	**Students use a partial-products algorithm and an improper-fraction algorithm to multiply a whole number and a mixed number.**	Incorporating and validating a variety of methods
8♦11	**Solve percent problems involving rural and urban populations.**	Connecting to everyday life

Vocabulary Development

The list below identifies the Key Vocabulary terms from this unit. The lesson in which each term is defined is indicated next to the term. Some of these terms or their homophones are used outside of mathematics. Consider adding other words as appropriate for developing understanding of the context of the lessons.

Lessons include suggestions for helping English language learners understand and develop vocabulary. For more information, see pages 17–19 of this handbook.

Key Vocabulary

area model (*model) **8◆6**

discount **8◆9**

horizontal **8◆5**

Quick Common Denominator (QCD) (*common) **8◆1**

unit fraction (*unit) **8◆10**

unit percent **8◆10**

vertical **8◆5**

- ★ Discuss the everyday and mathematical meanings of the words that are marked with an asterisk.
- ◆ As each word is introduced in the lesson, write the word on the board and discuss its meaning.
- ◆ List the words on a Math Word Wall for students to see. As each word is introduced in the lesson, add a picture next to the word on the Word Wall.
- ◆ Use the vocabulary words regularly when teaching lessons, and encourage students to use the words in their discussions.

Games

Below are suggested Unit 8 game adaptations. For more information about implementing games in a differentiated classroom, see pages 20–25 of this handbook.

Game: *Build-It*

Skill Practiced: Order fractions using benchmarks. [Number and Numeration Goal 6]

Modification	Purpose of Modification
Players make a deck of *Build-It* cards using these fractions: $\frac{1}{2}$, $\frac{1}{4}$, $\frac{2}{4}$, $\frac{3}{4}$, $\frac{1}{12}$, $\frac{2}{12}$, $\frac{3}{12}$, $\frac{4}{12}$, $\frac{5}{12}$, $\frac{6}{12}$, $\frac{7}{12}$, $\frac{8}{12}$, $\frac{9}{12}$, $\frac{10}{12}$, and $\frac{11}{12}$.	Students use benchmarks to order fractions with 2, 4, and 12 in the denominators. [Number and Numeration Goal 6]
Players get one "switch" at the end of the game where they can switch the position for two of their fraction cards.	Students compare and order fractions less than 1. [Number and Numeration Goal 6]

Game: *Mixed-Number Spin*

Skill Practiced: Estimate sums and differences of mixed numbers using benchmarks. [Operations and Computation Goals 4 and 6]

Modification	Purpose of Modification
Make a record sheet for each player that has only six addition number sentences. The sum of the two numbers in the number sentences should be set up as follows: $< 3, > 3, > 1, < 1, < 2, > \frac{1}{2}$.	Students estimate sums of fractions and mixed numbers. [Operations and Computation Goals 4 and 6]
Players find the actual sums and differences for their ten true number sentences. The final score is the sum of the solutions to the number sentences. The winner is the player with the largest overall sum.	Students add and subtract fractions and mixed numbers and find the sum for a multi-addend problem. [Operations and Computation Goal 4]

Game: *Fraction Action, Fraction Friction*

Skill Practiced: Estimate sums of fractions using benchmarks. [Operations and Computation Goals 4 and 6]

Modification	Purpose of Modification
Players use multiple copies of the twelfths fraction cards to play the game.	Students estimate sums of fractions with like denominators. [Operations and Computation Goals 4 and 6]
Players make a set of cards that includes fifths, eighths, and tenths.	Students add fractions with unlike denominators. [Operations and Computation Goals 4 and 6]

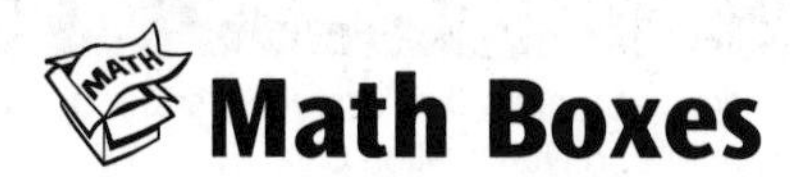

Math Boxes

Suggestions for using Math Boxes to meet individual needs begin on page 26 of this handbook. There are blank masters for Math Boxes on pages 136–141.

Using Part 3 of the Lessons

Use your professional judgment, along with assessment results, to determine whether the whole class, small groups, or individual students might benefit from these Unit 8 activities. Consider using the Part 3 Planning Master found on page 154 of this handbook to record your plans.

Readiness Activities

Lesson	Activity	Purpose of Activity
8•1	**Order fractions on a number line.**	Gain experience ordering fractions. [Number and Numeration Goal 6]
8•2	**Use an opposite-change algorithm to add mixed numbers.**	Explore mixed-number addition. [Operations and Computation Goal 4]
8•3	**Use a counting-up algorithm to subtract mixed numbers.**	Explore mixed-number subtraction. [Operations and Computation Goal 4]
8•4	**Use a flowchart to find common denominators before adding fractions.**	Explore fraction addition. [Operations and Computation Goal 4]
8•5	**Use fraction sticks to model equivalent fractions.**	Explore modeling equivalent fractional parts. [Number and Numeration Goal 5]
8•6	**Compare area models for multiplication of fractions.**	Explore the connection between the word *of* and multiplication. [Operations and Computation Goal 5]
8•7	**Rename whole numbers as fractions.**	Explore renaming whole numbers as fractions. [Number and Numeration Goal 5]
8•8	**Find common denominators for and order improper fractions.**	Explore ordering improper fractions. [Number and Numeration Goal 6]
8•9	**Use unit fractions to find the percent of a number.**	Explore how to find the percent of a number. [Number and Numeration Goal 2]
8•10	**Identify the errors in "fraction-of" and "percent-of" problems.**	Explore finding the fraction and the percent of a number. [Number and Numeration Goal 2]
8•11	**Use calculators to solve percent problems.**	Explore finding the percent of a number. [Number and Numeration Goal 2]
8•12	**Play *Build-It*.**	Explore comparing and ordering fractions and renaming mixed numbers as fractions. [Number and Numeration Goal 6]

English Language Learners Support Activities

Lesson	Activity	Purpose of Activity
8•1	**Add *quick common denominator* to the Math Word Bank.**	Make connections among and use visuals to represent terms. [Number and Numeration Goal 5]
8•5	**Add *horizontal* and *vertical* to the Math Word Bank.**	Make connections among and use visuals to represent terms. [Operations and Computation Goal 5]

Enrichment Activities

Lesson	Activity	Purpose of Activity
8•1	**Explore methods for finding least common multiples.**	Apply understanding of equivalent fractions. [Number and Numeration Goal 5]
8•3	**Explore patterns that result from adding and subtracting unit fractions.**	Apply understanding of addition and subtraction of fractions. [Operations and Computation Goal 4]
8•4	**Use calculators to explore fraction-to-decimal conversions.**	Apply understanding of equivalent fractions. [Number and Numeration Goal 5]
8•5	**Use a calculator to find fraction sums and square fractions.**	Explore computing with fractions. [Operations and Computation Goals 4 and 5]
8•7	**Use the Commutative Property to simplify fraction multiplication.**	Explore fraction multiplication and lowest terms. [Operations and Computation Goal 5; Patterns, Functions, and Algebra Goal 4]
8•9	**Find and calculate discounts for a variety of situations.**	Apply understanding of calculating percents. [Number and Numeration Goal 2]
8•10	**Identify appropriate and inappropriate methods to solve "fraction-of" and "percent-of" problems.**	Apply understanding of finding the fraction and the percent of a number. [Number and Numeration Goal 2]
8•11	**Make a line graph using food-consumption data.**	Apply understanding of interpreting data displays and organizing data. [Data and Chance Goals 1 and 2]
8•12	**Use calculators and fraction multiplication to find reciprocals of numbers.**	Explore the relationship between a number and its reciprocal. [Operations and Computation Goal 5]

Extra Practice Activities

Lesson	Activity	Purpose of Activity
8•2	**Play *Fraction Capture*.**	Practice finding equivalent fractions. [Number and Numeration Goal 5; Operations and Computation Goal 4]
8•2	**Solve mixed-number addition problems.**	Practice adding mixed numbers. [Operations and Computation Goal 4]
8•3	**Solve *5-Minute Math* problems involving adding mixed numbers.**	Practice adding mixed numbers. [Operations and Computation Goal 4]
8•4	**Solve *5-Minute Math* problems using calculators to add fractions.**	Practice adding fractions. [Operations and Computation Goal 4]
8•6	**Use an area model to multiply fractions.**	Practice fraction multiplication. [Operations and Computation Goal 5]
8•7	**Solve *5-Minute Math* problems involving multiplying fractions and whole numbers.**	Practice multiplying fractions and whole numbers. [Operations and Computation Goal 5]
8•8	**Play *Frac-Tac-Toe*.**	Practice converting among fractions, decimals, and percents. [Number and Numeration Goal 5]
8•12	**Divide fractions using a visual model.**	Practice dividing fractions. [Operations and Computation Goal 5]

Looking at Grade-Level Goals

Everyday Mathematics develops concepts and skills over time. Below is a chart showing where the Grade-Level Goals emphasized in this unit are addressed throughout the year. Use the chart to help you determine which Maintaining Concepts and Skills activities on page 105 to utilize to ensure that students continue working toward these Grade-Level Goals.

- ● Grade-Level Goal is taught.
- ◐ Grade-Level Goal is practiced and applied.
- ○ Grade-Level Goal is not a focus.

Grade-Level Goals Emphasized in Unit 8	Unit											
	1	2	3	4	5	6	7	8	9	10	11	12
[Number and Numeration Goal 2] Solve problems involving percents and discounts; describe and explain strategies used; identify the unit whole in situations involving fractions.	○	○	◐	○	●	◐	◐	●	○	○	◐	◐
[Number and Numeration Goal 5] Use numerical expressions to find and represent equivalent names for fractions, decimals, and percents; use and explain multiplication and division rules to find equivalent fractions and fractions in simplest form; convert between fractions and mixed numbers; convert between fractions, decimals, and percents.	◐	◐	○	◐	●	●	◐	●	◐	○	◐	◐
[Number and Numeration Goal 6] Compare and order rational numbers; use area models, benchmark fractions, and analyses of numerators and denominators to compare and order fractions and mixed numbers; describe strategies used to compare fractions and mixed numbers.	◐	◐	◐	◐	●	◐	●	●	◐	○	○	◐
[Operations and Computation Goal 4] Use mental arithmetic, paper-and-pencil algorithms and models, and calculators to solve problems involving the addition and subtraction of fractions and mixed numbers; describe the strategies used and explain how they work.	○	○	○	○	●	●	◐	●	◐	◐	○	○
[Operations and Computation Goal 5] Use area models, mental arithmetic, paper-and-pencil algorithms and models, and calculators to solve problems involving the multiplication of fractions and mixed numbers; use visual models, paper-and-pencil methods, and calculators to solve problems involving the division of fractions; describe the strategies used.	○	○	○	○	○	○	○	●	◐	○	○	◐
[Operations and Computation Goal 7] Use repeated addition, arrays, area, and scaling to model multiplication and division; use ratios expressed as words, fractions, percents, and with colons; solve problems involving ratios of parts of a set to the whole set.	●	◐	○	○	◐	○	◐	●	◐	●	○	◐

Maintaining Concepts and Skills

Those goals marked with an asterisk (*) are addressed in future units only as practice and application. Here are several suggestions for maintaining concepts and skills until goals are revisited.

Number and Numeration Goal 2*

- Have students play *Fraction Of*.
- Have students find errors in "fraction-of" and "percent-of" problems. See the Readiness activity in Lesson 8-10 for more information.
- Have students use calculators to find percents of numbers. See the Readiness activity in Lesson 8-11 for more information.
- Use the "What's My Rule?" master on page 146 of this handbook to create practice problems in which the rule is to find a fraction of the *in* number.

Number and Numeration Goal 5*

- Have students play *Fraction Capture* and *Frac-Tac-Toe.*
- Have students rename whole numbers as fractions and find common denominators. See the Readiness activity in Lesson 8-7 for more information.
- Use the Name-Collection Boxes master on page 147 of this handbook to create practice problems for fractions or mixed numbers.

Operations and Computation Goal 4*

- Have students play *Fraction Action, Fraction Friction* and record a number sentence to represent the final total for each round.
- Have students explore an alternate method for adding mixed numbers. See the Readiness activity in Lesson 8-2 for more information.

Operations and Computation Goal 5*

- Have students compare area models to reinforce the use of the word *of* to indicate multiplication. See the Readiness activity in Lesson 8-6 for more information.
- Use the "What's My Rule?" master on page 146 of this handbook to create practice problems in which the rule is to multiply by fractions or find the fraction of the *in* numbers.

Assessment

See page 112 in the *Assessment Handbook* for modifications to the written portion of the Unit 8 Progress Check.

Additionally, see pages 113–117 for modifications to the open-response task and selected student work samples.

Unit 9 Activities and Ideas for Differentiation

In this unit, students explore coordinate graphs, extend area concepts, and develop a formula for volume. This section summarizes opportunities for supporting multiple learning styles and ability levels. Use these suggestions to develop a differentiation plan for Unit 9.

Part 1 Activities That Support Differentiation

Below are examples of Unit 9 activities that highlight some of the general instructional strategies that are hallmarks of a differentiated classroom. These strategies will help you support, emphasize, and enhance lesson content to make sure all of your students are engaged in the mathematics at the highest possible level. For more information about general differentiation strategies that accommodate the diverse needs of today's classrooms, see the essay on pages 8–16 of this handbook.

Lesson	Activity	Strategy
9•2	**Students describe the changes in transformed sailboats.**	Talking about math
9•5	**Students use the rectangle method for finding area.**	Building on prior knowledge
9•6	**Students complete a table with base, height, and area and explore the relationships among these measures.**	Using organizational tools
9•7	**Students use longitude and latitude to locate places on Earth's coordinate grid.**	Connecting to everyday life
9•8	**On the board, record students' questions and answers about area and volume.**	Recording key ideas
9•10	**Demonstrate that there are 1,000 milliliters in one liter.**	Modeling concretely

Vocabulary Development

The list below identifies the Key Vocabulary terms from this unit. The lesson in which each term is defined is indicated next to the term. Some of these terms or their homophones are used outside of mathematics. Consider adding other words as appropriate for developing understanding of the context of the lessons.

Lessons include suggestions for helping English language learners understand and develop vocabulary. For more information, see pages 17–19 of this handbook.

Key Vocabulary

- altitude **9◆6**
- area **9◆4**
- Associative Property of Multiplication **9◆8**
- axes **9◆1**
- *†base **9◆4**
- base (of a rectangular prism) **9◆8**
- capacity **9◆10**
- *coordinate **9◆1**
- coordinate grid **9◆1**
- cubic centimeter **9◆10**
- cubic unit **9◆8**
- *cup (c) **9◆10**
- *face **9◆8**
- formula **9◆4**
- height **9◆4**
- height (of a rectangular prism) **9◆8**
- horizontal axis **9◆1**
- latitude **9◆7**
- liter (L) **9◆10**
- longitude **9◆7**
- milliliter (mL) **9◆10**
- opposite of a number **9◆3**
- ordered number pair (*ordered, *†pair) **9◆2**
- ordered pair of numbers **9◆1**
- origin **9◆1**
- perpendicular **9◆1**
- personal references **9◆5**
- *prism **9◆9**
- quart (qt) **9◆10**
- rectangle method **9◆5**
- rectangular prism **9◆8**
- reflection **9◆3**
- square units **9◆4**
- *translation **9◆3**
- variable **9◆4**
- vertical axis **9◆1**
- *volume **9◆8**
- volume of a container **9◆10**

* Discuss the everyday and mathematical meanings of the words that are marked with an asterisk.

† For words marked with a dagger, write the words and their homophones on the board. For example, *base* and *bass* and *pair, pear,* and *pare.* Discuss and clarify the meaning of each.

◆ As each word is introduced in the lesson, write the word on the board and discuss its meaning.

◆ List the words on a Math Word Wall for students to see. As each word is introduced in the lesson, add a picture next to the word on the Word Wall.

◆ Use the vocabulary words regularly when teaching lessons, and encourage students to use the words in their discussions.

Games

Below are suggested Unit 9 game adaptations. For more information about implementing games in a differentiated classroom, see pages 20–25 of this handbook.

Game: *Frac-Tac-Toe*

Skill Practiced: **Find equivalent names for fractions, decimals, and percents.** [Number and Numeration Goal 5]

Modification	Purpose of Modification
Players use the 2-4-5-10 percent gameboard and use only 4s and 10s in the denominator pile.	Students find equivalent names for percents with fourths and tenths. [Number and Numeration Goal 5]
Players add two 0s to the denominator pile. Since fractions cannot have a 0 in the denominator, the 0 card is like a wild card. When a 0 comes up, players can name any fraction they want and cover the percent for that fraction.	Students find equivalent names for fractions and percents. [Number and Numeration Goal 5]

Game: *Fraction Action, Fraction Friction*

Skill Practiced: **Estimate sums of fractions using benchmarks.** [Operations and Computation Goals 4 and 6]

Modification	Purpose of Modification
Provide players with fraction sticks that are divided into twelfths. For each card they draw, players shade the corresponding number of twelfths on their fraction sticks. Consider laminating the fraction sticks and using dry-erase markers so they are reusable.	Students add fractions on fraction sticks. [Operations and Computation Goals 4 and 6]
Players use the fractions on the Everything Math Deck to play the game. They may use a calculator to check their totals.	Students add fractions with unlike denominators. [Operations and Computation Goals 4 and 6]

Game: *Polygon Capture*

Skill Practiced: **Describe properties of polygons.** [Geometry Goal 2]

Modification	Purpose of Modification
Players use only Polygon Capture Pieces. Before beginning, they work together to make a list of polygon properties. On each turn, players turn over the top piece and win the piece if they can name a property of the polygon. Players get a bonus point if they can name a second property. Bonus points count as cards at the end of the game.	Students describe one property of a polygon. [Geometry Goal 2]
One time during the game, each player can call a "Free Turn" when they declare one "angle" and one "sides" property. They collect one face-up Polygon Capture Piece with both properties.	Students describe one side and one angle property of polygons. [Geometry Goal 2]

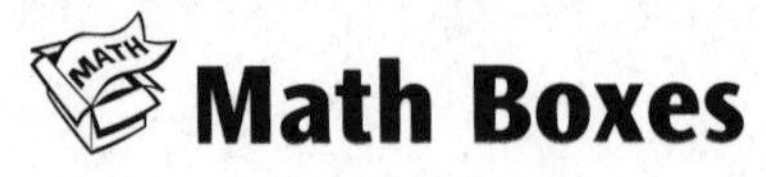

Math Boxes

Suggestions for using Math Boxes to meet individual needs begin on page 26 of this handbook. There are blank masters for Math Boxes on pages 136–141.

Using Part 3 of the Lessons

Use your professional judgment, along with assessment results, to determine whether the whole class, small groups, or individual students might benefit from these Unit 9 activities. Consider using the Part 3 Planning Master found on page 154 of this handbook to record your plans.

Readiness Activities

Lesson	Activity	Purpose of Activity
9•1	**Use ordered pairs of numbers to identify locations on a map.**	Gain experience with coordinate grids. [Measurement and Reference Frames Goal 4]
9•2	**Plot and label points between whole numbers on a number line.**	Explore locating decimals on a number line. [Number and Numeration Goal 6]
9•3	**Build a coordinate grid using tape and metersticks.**	Explore coordinate grids. [Measurement and Reference Frames Goal 4]
9•4	**Compare the perimeters and areas for rectangles using centimeter grids.**	Explore relationships between perimeter and area. [Measurement and Reference Frames Goal 2]
9•5	**Construct rectangles by arranging triangles.**	Explore relationships between triangles and rectangles. [Geometry Goal 2]
9•6	**Transform parallelograms and multiple triangles into rectangles to derive the area formulas.**	Explore how area formulas for parallelograms and triangles are derived. [Measurement and Reference Frames Goal 2]
9•8	**Determine which diagrams can be folded into cubes.**	Explore properties of geometric solids. [Geometry Goal 2]
9•9	**Construct a triangular prism from a net.**	Explore properties of triangular prisms. [Geometry Goal 2]
9•10	**Compare the capacity of containers.**	Explore the concept of capacity. [Measurement and Reference Frames Goal 3]

English Language Learners Support Activities

Lesson	Activity	Purpose of Activity
9•1	**Add *horizontal axis* and *vertical axis* to the Math Word Bank.**	Make connections among and use visuals to represent terms. [Measurement and Reference Frames Goal 4]
9•2	**Plot and label coordinates on a grid and label *horizontal axis, ordered number pair,* and *vertical axis.***	Clarify the mathematical uses of the terms. [Measurement and Reference Frames Goal 4]
9•3	**Add *reflection* and *translation* to the Math Word Bank.**	Make connections among and use visuals to represent terms. [Geometry Goal 3]
9•4	**Add *length, height, base,* and *width* to the Math Word Bank.**	Make connections among and use visuals to represent terms. [Measurement and Reference Frames Goal 2]
9•10	**Explore prefixes used in *metric measurement.***	Clarify mathematical and everyday uses of prefixes. [Measurement and Reference Frames Goal 1]

Enrichment Activities

Lesson	Activity	Purpose of Activity
9•1	Use grids to compare and analyze diagonal and gridline distances.	Apply understanding of coordinate grids. [Measurement and Reference Frames Goal 4]
9•2	Create rules to change the values of ordered number pairs.	Apply understanding of transforming figures on a coordinate grid. [Geometry Goal 3]
9•3	Record coordinates and plot reflections for letters of the alphabet.	Apply understanding of transformations on a coordinate grid. [Measurement and Reference Frames Goal 4; Geometry Goal 3]
9•4	Compare the perimeters and areas of irregular polygons.	Apply understanding of perimeter and area. [Measurement and Reference Frames Goal 2]
9•6	Calculate the area of a nonrectangular path on a grid.	Apply understanding of the rectangle method and formulas for finding area. [Measurement and Reference Frames Goal 2]
9•7	Estimate the ground-level area of the school.	Apply understanding of sampling as an estimation strategy. [Operations and Computation Goal 6]
9•8	Find the volume of one stick-on note.	Apply understanding of how to find the volume of a rectangular prism. [Measurement and Reference Frames Goal 2]
9•9	Build nets for prisms.	Explore properties of geometric solids. [Geometry Goal 2]

Extra Practice Activities

Lesson	Activity	Purpose of Activity
9•2	Draw pictures on a coordinate grid and list the plotted points.	Practice plotting points on a coordinate grid. [Measurement and Reference Frames Goal 4]
9•4	Solve *5-Minute Math* problems involving the areas of rectangles.	Practice calculating the areas of rectangles. [Measurement and Reference Frames Goal 2]
9•4	Use tiling and formulas to find the area of rectangles.	Practice finding the area of rectangles using tiling and a formula. [Measurement and Reference Frames Goal 2]
9•5	Read *Spaghetti and Meatballs for All* and solve problems like those in the story.	Practice calculating perimeter and area. [Measurement and Reference Frames Goal 2]
9•6	Solve *5-Minute Math* problems involving finding the areas of triangles.	Practice calculating the areas of triangles. [Measurement and Reference Frames Goal 2]
9•7	Use formulas to find the areas of nonrectangular figures.	Practice using formulas to find the areas of figures. [Measurement and Reference Frames Goal 2]
9•8	Solve *5-Minute Math* problems involving finding the volumes of prisms.	Practice calculating the volumes of prisms. [Measurement and Reference Frames Goal 2]
9•10	Read and discuss *Room for Ripley.*	Explore volume. [Measurement and Reference Frames Goal 2]

Looking at Grade-Level Goals

Everyday Mathematics develops concepts and skills over time. Below is a chart showing where the Grade-Level Goals emphasized in this unit are addressed throughout the year. Use the chart to help you determine which Maintaining Concepts and Skills activities on page 112 to utilize to ensure that students continue working toward these Grade-Level Goals.

- ● Grade-Level Goal is taught.
- ◉ Grade-Level Goal is practiced and applied.
- ○ Grade-Level Goal is not a focus.

Grade-Level Goals Emphasized in Unit 9	Unit											
	1	2	3	4	5	6	7	8	9	10	11	12
[Measurement and Reference Frames Goal 2] Describe and use strategies to find the perimeter of polygons and the area of circles; choose and use appropriate methods, including formulas, to find the areas of rectangles, parallelograms, and triangles, and the volume of a prism; define *pi* as the ratio of a circle's circumference to its diameter.	○	◉	◉	○	○	○	○	◉	●	●	◉	◉
[Measurement and Reference Frames Goal 3] Describe relationships among U.S. customary units of measure and among metric units of measure.	○	◉	○	○	◉	○	○	○	●	◉	●	○
[Measurement and Reference Frames Goal 4] Use ordered pairs of numbers to name, locate, and plot points in all four quadrants of a coordinate grid.	○	○	○	○	○	○	○	◉	●	◉	○	◉
[Geometry Goal 2] Describe, compare, and classify plane and solid figures using appropriate geometric terms; identify congruent figures and describe their properties.	○	◉	●	○	○	○	○	◉	◉	◉	◉	○
[Geometry Goal 3] Identify, describe, and sketch examples of reflections, translations, and rotations.	○	○	◉	○	○	○	○	○	●	○	○	○

Maintaining Concepts and Skills

Some of the goals addressed in this unit will be addressed again in later units. Those goals marked with an asterisk (*) are addressed in future units only as practice and application. Here are several suggestions for maintaining concepts and skills until goals are revisited.

Measurement and Reference Frames Goal 2

- Have students compare the perimeters and areas of rectangles. See the Readiness activity in Lesson 9-2 for more information.
- Have students explore the area formulas for parallelograms and triangles. See the Readiness activity in Lesson 9-6 for more information.

Measurement and Reference Frames Goal 3

- Use the "What's My Rule?" master on page 146 of this handbook to create practice problems in which the rule is to convert to an equivalent measure. For example, the *in* number is the number of inches and the *out* number is the number of feet.

Measurement and Reference Frames Goal 4*

- Have students play *Hidden Treasure.*
- Have students find locations on a map. See the Readiness activity in Lesson 9-1 for more information.
- Have students build a coordinate grid. See the Readiness activity in Lesson 9-3 for more information.

Geometry Goal 3*

- Have students use pattern blocks to make tessellating patterns.
- Have students identify tessellating patterns in the classroom, for example, the floor tiles or ceiling tiles. Consider having students also identify non-tessellating patterns, for example, clothing prints.

Assessment

See page 120 in the *Assessment Handbook* for modifications to the written portion of the Unit 9 Progress Check.

Additionally, see pages 121–125 for modifications to the open-response task and selected student work samples.

Activities and Ideas for Differentiation

In this unit, students solve equations using a pan-balance metaphor and explore representing relationships as algebraic expressions containing variables. This section summarizes opportunities for supporting multiple learning styles and ability levels. Use these suggestions to develop a differentiation plan for Unit 10.

Part 1 Activities That Support Differentiation

Below are examples of Unit 10 activities that highlight some of the general instructional strategies that are hallmarks of a differentiated classroom. These strategies will help you support, emphasize, and enhance lesson content to make sure all of your students are engaged in the mathematics at the highest possible level. For more information about general differentiation strategies that accommodate the diverse needs of today's classrooms, see the essay on pages 8–16 of this handbook.

Lesson	Activity	Strategy
10•1	**Demonstrate how to use a pan balance to solve equations.**	Building on prior knowledge
10•3	**Construct a rule table for algebraic expressions.**	Using organizational tools
10•4	**Students share their strategies for solving a rate problem.**	Incorporating and validating a variety of methods
10•5	**Students construct tables and graphs and write rules for Old Faithful's eruptions.**	Connecting to everyday life
10•7	**Students discuss and compare line graphs for different data sets.**	Talking about math
10•9	**Students count squares to find the area of a circle.**	Modeling concretely

Vocabulary Development

The list below identifies the Key Vocabulary terms from this unit. The lesson in which each term is defined is indicated next to the term. Some of these terms or their homophones are used outside of mathematics. Consider adding other words as appropriate for developing understanding of the context of the lessons.

Lessons include suggestions for helping English language learners understand and develop vocabulary. For more information, see pages 17–19 of this handbook.

Key Vocabulary	
algebraic expression (*expression) **10◆3**	pan balance (*balance) **10◆1**
circumference **10◆8**	†pi (π) **10◆8**
*coordinates **10◆6**	predict **10◆5**
diameter **10◆8**	radius **10◆8**
*formula **10◆4**	*rate **10◆4**
geyser **10◆5**	rate of speed **10◆4**
line graph **10◆4**	ratio **10◆8**
mystery graph **10◆7**	ratio comparison **10◆8**
ordered number pairs (*ordered, *†pair) **10◆6**	variable **10◆4**

* Discuss the everyday and mathematical meanings of the words that are marked with an asterisk.

† For words marked with a dagger, write the words and their homophones on the board. For example, *pair, pare,* and *pear* and *pi* and *pie.* Discuss and clarify the meaning of each.

- ◆ As each word is introduced in the lesson, write the word on the board and discuss its meaning.
- ◆ List the words on a Math Word Wall for students to see. As each word is introduced in the lesson, add a picture next to the word on the Word Wall.
- ◆ Use the vocabulary words regularly when teaching lessons, and encourage students to use the words in their discussions.

Games

Below are suggested Unit 10 game adaptations. For more information about implementing games in a differentiated classroom, see pages 20–25 of this handbook.

Game: *First to 100*

Skill Practiced: Solve open number sentences. [Patterns, Functions, and Algebra Goal 2]

Modification	Purpose of Modification
Players use the sum of their dice instead of the product. Play ends after ten rounds. The player with the higher score wins.	Students solve open number sentences. [Patterns, Functions, and Algebra Goal 2]
Players make a 2-digit number from the roll of the dice. If the final answer is a multiple of 3 or of 4, they get ten bonus points. Play ends after ten rounds. The player with the highest score wins.	Students use place-value concepts to make and solve open number sentences. [Patterns, Functions, and Algebra Goal 2]

Game: *Name That Number*

Skill Practiced: Find equivalent names for numbers using multiple operations. [Number and Numeration Goal 4; Patterns, Functions, and Algebra Goal 3]

Modification	Purpose of Modification
Players draw six cards instead of five to increase their options.	Students find equivalent names for numbers using multiple operations. [Number and Numeration Goal 4; Patterns, Functions, and Algebra Goal 3]
Players represent the target number using at least three operations in each solution.	Students find equivalent names for numbers using multiple operations. [Number and Numeration Goal 4; Patterns, Functions, and Algebra Goal 3]

Game: *Fraction Spin*

Skill Practiced: Estimate sums and differences of fractions. [Operations and Computation Goals 4 and 6]

Modification	Purpose of Modification
Players change the fractions on the spinner from thirds to fourths and from sixths to eighths.	Students estimate sums and differences of fractions with 2, 4, and 8 in the denominators. [Operations and Computation Goals 4 and 6]
Players add a third addend to three of the addition number sentences.	Students estimate sums and differences of fractions and find the sum for multi-addend problems. [Operations and Computation Goals 4 and 6]

Math Boxes

Suggestions for using Math Boxes to meet individual needs begin on page 26 of this handbook. There are blank masters for Math Boxes on pages 136–141.

Using Part 3 of the Lessons

Use your professional judgment, along with assessment results, to determine whether the whole class, small groups, or individual students might benefit from these Unit 10 activities. Consider using the Part 3 Planning Master found on page 154 of this handbook to record your plans.

Readiness Activities

Lesson	Activity	Purpose of Activity
10•1	**Use small objects and a pan balance to compare weights of various objects.**	Gain experience with a pan-balance model of equality. [Patterns, Functions, and Algebra Goal 2]
10•2	**Solve logic puzzles related to pan-balance problems.**	Gain experience with deductive reasoning required to solve linear equations. [Patterns, Functions, and Algebra Goal 2]
10•3	**Write and solve "What's My Rule?" problems.**	Gain experience using patterns in tables to solve problems. [Patterns, Functions, and Algebra Goal 1]
10•6	**Identify the relationship between parts of a table and parts of a graph.**	Gain experience constructing graphs from table data. [Data and Chance Goal 1; Patterns, Functions, and Algebra Goal 1]
10•7	**Match silhouettes to objects or activities.**	Explore shape outlines to prepare for interpreting mystery graphs. [Patterns, Functions, and Algebra Goal 1]
10•7	**Identify and describe errors in graphs.**	Gain experience analyzing graphs. [Data and Chance Goal 2; Patterns, Functions, and Algebra Goal 1]
10•8	**Read and discuss *The Librarian Who Measured the Earth.***	Explore the relationship between circumference and diameter. [Measurement and Reference Frames Goal 2]

English Language Learners Support Activities

Lesson	Activity	Purpose of Activity
10•3	**Add *algebraic expression* to the Math Word Bank.**	Make connections among and use visuals to represent terms. [Patterns, Functions, and Algebra Goal 2]
10•4	**Describe situations involving *rates* and *ratios* displayed in the Rates and Ratios Museum.**	Make connections between mathematics and everyday life; discuss new mathematical ideas. [Operations and Computation Goal 7]
10•8	**Add *rate, ratio,* and *ratio comparison* to the Math Word Bank.**	Make connections among and use visuals to represent terms. [Operations and Computation Goal 7]

Enrichment Activities

Lesson	Activity	Purpose of Activity
10•1	**Describe how to use a pan balance to solve a penny riddle.**	Apply understanding of the pan-balance model of equality. [Patterns, Functions, and Algebra Goal 2]
10•2	**Use sandglasses to solve a problem about time intervals.**	Explore solving a problem with two unknowns. [Patterns, Functions, and Algebra Goal 2]
10•3	**Complete tables, write rules, and graph ordered pairs to identify relationships.**	Apply understanding of rules and patterns to patterns with two rules. [Patterns, Functions, and Algebra Goal 1]
10•4	**Write rate formulas and use them to solve rate problems.**	Apply understanding of rate formulas. [Operations and Computation Goal 7; Patterns, Functions, and Algebra Goal 1]
10•5	**Complete and graph values from "What's My Rule?" tables.**	Apply understanding of representing functions with graphs. [Patterns, Functions, and Algebra Goal 1]
10•6	**Graph several sets of race-result data on the same grid.**	Apply understanding of constructing graphs from table data. [Data and Chance Goal 1; Patterns, Functions, and Algebra Goal 1]
10•7	**Make data tables from graphs and write the rules for each.**	Apply understanding of the relationship between line graphs and data tables. [Data and Chance Goal 1; Patterns, Functions, and Algebra Goal 1]
10•8	**Use scissors to reconfigure a rectangular piece of paper into a large circle.**	Apply understanding of circumference. [Measurement and Reference Frames Goal 2]
10•9	**Cut a circle into sectors and arrange the sectors into a parallelogram to explore a model for the formula $A = \pi r^2$.**	Apply understanding of area formulas. [Measurement and Reference Frames Goal 2]

Extra Practice Activities

Lesson	Activity	Purpose of Activity
10•1	**Weigh pennies with a pan balance to determine the year the weight of a penny changed.**	Practice using a pan balance to solve problems. [Patterns, Functions, and Algebra Goal 2]
10•3	**Write expressions for problems.**	Practice writing algebraic expressions. [Patterns, Functions, and Algebra Goal 2]
10•4	**Generate and graph values in "What's My Rule?" tables.**	Practice graphing ordered pairs of numbers. [Patterns, Functions, and Algebra Goal 1; Data and Chance Goal 1]
10•4	**Complete a table and graph values to identify relationships.**	Practice analyzing data through the use of tables and graphs. [Patterns, Functions, and Algebra Goal 1]
10•5	**Solve *5-Minute Math* problems involving graphing data.**	Practice graphing data. [Data and Chance Goal 1]
10•6	**Complete a table of values to graph and analyze data.**	Practice analyzing data with two rules. [Patterns, Functions, and Algebra Goal 1]
10•7	**Solve *5-Minute Math* problems involving graphing data landmarks.**	Practice identifying data landmarks. [Data and Chance Goal 2]
10•9	**Use a calculator to find the circumference and area of circles.**	Practice finding circumference and area of circles. [Measurement and Reference Frames Goal 2]

Looking at Grade-Level Goals

Everyday Mathematics develops concepts and skills over time. Below is a chart showing where the Grade-Level Goals emphasized in this unit are addressed throughout the year. Use the chart to help you determine which Maintaining Concepts and Skills activities on page 119 to utilize to ensure that students continue working toward these Grade-Level Goals.

- ● Grade-Level Goal is taught.
- ◍ Grade-Level Goal is practiced and applied.
- ○ Grade-Level Goal is not a focus.

Grade-Level Goals Emphasized in Unit 10	Unit											
	1	2	3	4	5	6	7	8	9	10	11	12
[Operations and Computation Goal 1] Use manipulatives, mental arithmetic, paper-and-pencil algorithms and models, and calculators to solve problems involving the addition and subtraction of whole numbers, decimals, and signed numbers; describe the strategies used and explain how they work.	◍	●	◍	◍	◍	◍	●	◍	◍	◍	◍	○
[Operations and Computation Goal 7] Use repeated addition, arrays, area, and scaling to model multiplication and division; use ratios expressed as words, fractions, percents, and with colons; solve problems involving ratios of parts of a set to the whole set.	●	◍	○	○	◍	○	◍	●	◍	●	○	◍
[Data and Chance Goal 1] Collect and organize data or use given data to create graphic displays with reasonable titles, labels, keys, and intervals.	○	◍	○	○	●	●	◍	◍	○	◍	◍	○
[Data and Chance Goal 2] Use the maximum, minimum, range, median, mode, and mean and graphs to ask and answer questions, draw conclusions, and make predictions.	◍	●	●	◍	◍	●	◍	○	◍	◍	○	○
[Measurement and Reference Frames Goal 2] Describe and use strategies to find the perimeter of polygons and the area of circles; choose and use appropriate methods, including formulas, to find the areas of rectangles, parallelograms, and triangles, and the volume of a prism; define *pi* as the ratio of a circle's circumference to its diameter.	○	◍	◍	○	○	○	○	◍	●	●	◍	◍
[Patterns, Functions, and Algebra Goal 1] Extend, describe, and create numeric patterns; describe rules for patterns and use them to solve problems; write rules for functions involving the four basic arithmetic operations; represent functions using words, symbols, tables, and graphs and use those representations to solve problems.	●	○	◍	○	○	○	●	●	◍	●	○	◍
[Patterns, Functions, and Algebra Goal 2] Determine whether number sentences are true or false; solve open number sentences and explain the solutions; use a letter variable to write an open sentence to model a number story; use a pan-balance model to solve linear equations in one unknown.	○	◍	◍	●	◍	◍	●	◍	○	●	◍	◍

Maintaining Concepts and Skills

Those goals marked with an asterisk (*) are addressed in future units only as practice and application. Here are several suggestions for maintaining concepts and skills until goals are revisited.

Operations and Computation Goal 1*

- Have students play *Addition* or *Subtraction Top-It, Subtraction Target Practice,* and *Top-It with Positive and Negative Numbers.*
- Use Frames-and-Arrows masters A and B on pages 144 and 145 of this handbook to create practice problems with rules that involve addition or subtraction.

Data and Chance Goal 2*

- Have students make tally charts and graphs for survey data, for example, height, time it takes to get to school, or time it takes to do homework. Then have them find the data landmarks for the survey data sets.

Measurement and Reference Frames Goal 2*

- Read *The Librarian Who Measured the Earth* and discuss the relationship between circumference and diameter.

Patterns, Functions, and Algebra Goal 1*

- Have students identify and describe errors in graphs based on the patterns they see. See the Readiness activity in Lesson 10-7 for more information.
- Use the "What's My Rule?" master on page 146 of this handbook to create practice problems.

Patterns, Functions, and Algebra Goal 2*

- Have students play *First to 100.*
- Have students use a pan balance to explore equivalencies. See the Readiness activity in Lesson 10-1 for more information.

Assessment

See page 128 in the *Assessment Handbook* for modifications to the written portion of the Unit 10 Progress Check.

Additionally, see pages 129–133 for modifications to the open-response task and selected student work samples.

Activities and Ideas for Differentiation

In this unit, students review properties of 3-dimensional shapes and explore volume formulas for prisms, pyramids, cylinders, and cones. This section summarizes opportunities for supporting multiple learning styles and ability levels. Use these suggestions to develop a differentiation plan for Unit 11.

Part 1 Activities That Support Differentiation

Below are examples of Unit 11 activities that highlight some of the general instructional strategies that are hallmarks of a differentiated classroom. These strategies will help you support, emphasize, and enhance lesson content to make sure all of your students are engaged in the mathematics at the highest possible level. For more information about general differentiation strategies that accommodate the diverse needs of today's classrooms, see the essay on pages 8–16 of this handbook.

Lesson	Activity	Strategy
11•1	**Students name objects that are shaped like geometric solids.**	Connecting to everyday life
11•2	**Draw a Venn diagram to compare the properties of prisms and pyramids.**	Using organizational tools
11•3	**On the board, record important information about finding the area of a circle.**	Recording key ideas
11•4	**Demonstrate the relationship between the volume of a pyramid and a prism with the same base and height.**	Modeling concretely
11•5	**Demonstrate how to find the volume of irregular solids using a displacement method.**	Modeling concretely
11•6	**Students describe a visual model for comparing measures of capacity.**	Talking about math

Vocabulary Development

The list below identifies the Key Vocabulary terms from this unit. The lesson in which each term is defined is indicated next to the term. Some of these terms or their homophones are used outside of mathematics. Consider adding other words as appropriate for developing understanding of the context of the lessons.

Lessons include suggestions for helping English language learners understand and develop vocabulary. For more information, see pages 17–19 of this handbook.

Key Vocabulary

apex **11◆2**

*†base **11◆2**

calibrate **11◆5**

cone **11◆1**

cylinder **11◆1**

displacement **11◆5**

*edge **11◆1**

face **11◆1**

geometric solid **11◆1**

polyhedron (regular) **11◆1**

*prism **11◆1**

pyramid **11◆1**

sphere **11◆1**

surface **11◆1**

surface area **11◆7**

vertex (vertices or vertexes) **11◆1**

* Discuss the everyday and mathematical meanings of the words that are marked with an asterisk.

† For the word marked with a dagger, write *base* and its homophone *bass* on the board. Discuss and clarify the meaning of each.

◆ As each word is introduced in the lesson, write the word on the board and discuss its meaning.

◆ List the words on a Math Word Wall for students to see. As each word is introduced in the lesson, add a picture next to the word on the Word Wall.

◆ Use the vocabulary words regularly when teaching lessons, and encourage students to use the words in their discussions.

Games

Below are suggested Unit 11 game adaptations. For more information about implementing games in a differentiated classroom, see pages 20–25 of this handbook.

Game: *3-D Shape Sort*

Skill Practiced: Identify and describe properties of 3-D shapes. [Geometry Goal 2]

Modification	Purpose of Modification
Players use 3-D figures instead of the Shape Cards. They collect shapes in the classroom or use a set of 3-D solids.	Students identify and describe properties of 3-D shapes. [Geometry Goal 2]
Mix two sets of Property Cards together (both vertices/edges and sides/faces cards). Players draw two Property Cards and use one or two to make a match. They collect only one shape per turn. Players "win" Property Card(s) they use and the Shape Card they collect. Any remaining cards can be used in the next round. The player with the most cards wins.	Students identify and describe multiple properties of 3-D shapes. [Geometry Goal 2]

Game: *Rugs and Fences*

Skill Practiced: Calculate the area and perimeters of polygons. [Measurement and Reference Frames Goal 2]

Modification	Purpose of Modification
Players calculate either area or perimeter consistently on every turn. The "choice" cards become "double" cards. If the player makes the designated calculation, the score is doubled.	Students calculate either the areas or perimeters of polygons. [Measurement and Reference Frames Goal 2]
Players calculate both the area and the perimeter for each polygon. Their score is the combined total. The "choice" cards become "double" cards. If the player makes the correct calculation, the score is doubled.	Students calculate both the areas and perimeters of polygons. [Measurement and Reference Frames Goal 2]

Game: *Name That Number*

Skill Practiced: Students write equivalent names for numbers using multiple operations. [Number and Numeration Goal 4; Patterns, Functions, and Algebra Goal 3]

Modification	Purpose of Modification
Players use only addition and subtraction to make the target number.	Students write equivalent names for numbers using addition and subtraction. [Number and Numeration Goal 4; Patterns, Functions, and Algebra Goal 3]
Players draw two cards to make the target number. They can make the target a 2-digit number or a fraction. They may multiply individual digit cards in their hand by any power of 10. For example, if they have a 2 in their hand, they can use it as a 0.2, 2, 20, and so on.	Students write equivalent names for rational numbers using multiple operations. [Number and Numeration Goal 4; Patterns, Functions, and Algebra Goal 3]

Math Boxes

Suggestions for using Math Boxes to meet individual needs begin on page 26 of this handbook. There are blank masters for Math Boxes on pages 136–141.

Using Part 3 of the Lessons

Use your professional judgment, along with assessment results, to determine whether the whole class, small groups, or individual students might benefit from these Unit 11 activities. Consider using the Part 3 Planning Master found on page 154 of this handbook to record your plans.

Readiness Activities

Lesson	Activity	Purpose of Activity
11•2	**Investigate and discuss ideas from *Flatland.***	Explore the concept of dimension in geometry. [Geometry Goal 2]
11•3	**Construct paper cylinders and maximize/minimize their volume.**	Explore comparing the volumes of cylinders. [Measurement and Reference Frames Goal 2]
11•4	**Find and compare the areas of concentric circles.**	Explore the relationship between diameters and areas of circles. [Measurement and Reference Frames Goal 2]
11•5	**Read and discuss *Who Sank the Boat?***	Explore the concept of displacement. [Measurement and Reference Frames Goal 2]
11•6	**Measure, weigh, and compare cups, ounces, and pounds.**	Explore converting among customary units of capacity and weight. [Measurement and Reference Frames Goal 3]

English Language Learners Support Activities

Lesson	Activity	Purpose of Activity
11•1	**Describe attributes of *prisms, pyramids, cylinders, cones,* and *spheres.***	Clarify the mathematical uses of the terms. [Geometry Goal 2]
11•4	**Add *volume of a cone* to the Math Word Bank.**	Make connections among and use visuals to represent terms. [Measurement and Reference Frames Goal 2]
11•6	**Add *volume, weight,* and *capacity* to the Math Word Bank.**	Make connections among and use visuals to represent terms. [Measurement and Reference Frames Goals 1 and 2]

Enrichment Activities

Lesson	Activity	Purpose of Activity
11•1	**Explore the relationship between the number of faces, vertices, and edges in polyhedrons and verify Euler's theorem.**	Apply understanding of the properties of geometric solids. [Geometry Goal 2]
11•2	Construct an octahedron and a truncated octahedron and explore other truncated polyhedrons.	Explore properties of geometric solids. [Geometry Goal 2]
11•4	**Modify the distance between concentric circles to enlarge or shrink the regions.**	Apply understanding of area. [Measurement and Reference Frames Goal 2]
11•5	Use the principle of displacement to find volume.	Apply understanding of displacement. [Measurement and Reference Frames Goal 2]
11•6	**Add and subtract measurements.**	Apply understanding of measurement conversions. [Measurement and Reference Frames Goal 3]
11•7	For a given volume, find the dimensions of the rectangular prism with the least surface area.	Apply understanding of volume and surface area. [Measurement and Reference Frames Goal 2]

Extra Practice Activities

Lesson	Activity	Purpose of Activity
11•1	**Build models for a rectangular prism and an octahedron from patterns.**	Practice making models of geometric solids. [Geometry Goal 2]
11•3	Calculate the volumes of cylindrical classroom objects.	Practice calculating the volumes of cylinders. [Measurement and Reference Frames Goal 2]
11•3	**Solve *5-Minute Math* problems involving geometric solids.**	Practice identifying geometric solids. [Geometry Goal 2]
11•4	Solve *5-Minute Math* problems involving the properties of geometric solids.	Practice identifying the properties of geometric solids. [Geometry Goal 2]
11•6	**Convert customary measures of capacity to other measures of capacity or weight.**	Practice converting among customary units of capacity and units of weight. [Measurement and Reference Frames Goal 3]
11•7	Find the area, surface area, and volume of geometric solids.	Practice calculating area, surface area, and volume. [Measurement and Reference Frames Goal 2]

Looking at Grade-Level Goals

Everyday Mathematics develops concepts and skills over time. Below is a chart showing where the Grade-Level Goals emphasized in this unit are addressed throughout the year. Use the chart to help you determine which Maintaining Concepts and Skills activities on page 126 to utilize to ensure that students continue working toward these Grade-Level Goals.

● Grade-Level Goal is taught.
◍ Grade-Level Goal is practiced and applied.
○ Grade-Level Goal is not a focus.

Grade-Level Goals Emphasized in Unit 11	Unit											
	1	2	3	4	5	6	7	8	9	10	11	12
[Data and Chance Goal 1] Collect and organize data or use given data to create graphic displays with reasonable titles, labels, keys, and intervals.	○	◍	○	○	●	●	◍	◍	○	◍	◍	○
[Measurement and Reference Frames Goal 2] Describe and use strategies to find the perimeter of polygons and the area of circles; choose and use appropriate methods, including formulas, to find the areas of rectangles, parallelograms, and triangles, and the volume of a prism; define *pi* as the ratio of a circle's circumference to its diameter.	○	◍	◍	○	○	○	○	◍	●	●	◍	◍
[Measurement and Reference Frames Goal 3] Describe relationships among U.S. customary units of measure and among metric units of measure.	○	◍	○	○	◍	○	○	○	●	◍	●	○
[Geometry Goal 2] Describe, compare, and classify plane and solid figures using appropriate geometric terms; identify congruent figures and describe their properties.	○	◍	●	○	○	○	○	◍	◍	◍	◍	○

Maintaining Concepts and Skills

Those goals marked with an asterisk (*) are addressed in future units only as practice and application. Here are several suggestions for maintaining concepts and skills until goals are revisited.

Data and Chance Goal 1*

- Have students routinely collect and organize survey data, for example, favorite pet, favorite color, methods for traveling to school, and so on. These can be displayed in a Surveys and Survey Data Museum. Have students compare and discuss features of the different data sets.

Measurement and Reference Frames Goal 2*

- Have students play *Rugs and Fences.*
- Have students compare the volume of cylinders. See the Readiness activity in Lesson 11-3 for more information.
- Have students discuss displacement. See the Readiness activity in Lesson 11-5 for more information.

Measurement and Reference Frames Goal 3*

- Have students explore equivalencies between units of capacity and weight. See the Readiness activity in Lesson 11-6 for more information.
- Use the "What's My Rule?" master on page 146 of this handbook to create practice problems for equivalent measures. For example, the *in* number is the number of hours, and the *out* number is the number of minutes.

Geometry Goal 2*

- Have students play *3-D Shape Sort.*
- Have students collect and label geometric solids. Use Venn Diagram masters A and B on pages 149 and 150 of this handbook to compare and contrast properties of the solids.

Assessment

See page 136 in the *Assessment Handbook* for modifications to the written portion of the Unit 11 Progress Check.

Additionally, see pages 137–141 for modifications to the open-response task and selected student work samples.

Activities and Ideas for Differentiation

In this unit, students explore factor trees to find prime factorization and tree diagrams to represent and count combinations of choices, and review the uses of ratios and rates. This section summarizes opportunities for supporting multiple learning styles and ability levels. Use these suggestions to develop a differentiation plan for Unit 12.

Part 1 Activities That Support Differentiation

Below are examples of Unit 12 activities that highlight some of the general instructional strategies that are hallmarks of a differentiated classroom. These strategies will help you support, emphasize, and enhance lesson content to make sure all of your students are engaged in the mathematics at the highest possible level. For more information about general differentiation strategies that accommodate the diverse needs of today's classrooms, see the essay on pages 8–16 of this handbook.

Lesson	Activity	Strategy
12•1	**Students use factor trees to find prime factorizations and to identify greatest common factors.**	Using organizational tools
12•2	**Display all combinations of entrances and exits for a stadium.**	Using a visual reference
12•3	**Students use ratios to examine trends.**	Connecting to everyday life
12•4	**Students use tiles to model ratio problems.**	Modeling concretely
12•5	**Students explain and illustrate their solution strategies for ratio problems.**	Talking about math
12•7	**Students collect and analyze data about their heart rates.**	Connecting to everyday life

Vocabulary Development

The list below identifies the Key Vocabulary terms from this unit. The lesson in which each term is defined is indicated next to the term. Some of these terms or their homophones are used outside of mathematics. Consider adding other words as appropriate for developing understanding of the context of the lessons.

Lessons include suggestions for helping English language learners understand and develop vocabulary. For more information, see pages 17–19 of this handbook.

Key Vocabulary

carbon dioxide (*carbon) **12◆8**

cardiac output **12◆8**

common factor (*common, *factor) **12◆1**

equally likely **12◆2**

factor tree (*tree) **12◆1**

greatest common factor (*common) **12◆1**

heart rate **12◆6**

least common multiple (†least) 12◆1

magnitude **12◆3**

Multiplication Counting Principle (†principle) **12◆2**

nutrients **12◆8**

oxygen **12◆8**

prime factorization (*prime) **12◆1**

probability **12◆2**

*profile **12◆7**

pulse **12◆6**

pulse rate **12◆6**

*rate **12◆7**

ratio **12◆3**

ratio comparison **12◆3**

target heart rate (*target) **12◆7**

tree diagram **12◆2**

★ Discuss the everyday and mathematical meanings of the words that are marked with an asterisk.

† For words marked with a dagger, write the words and their homophones on the board. For example, *least* and *leased* and *principle* and *principal.* Discuss and clarify the meaning of each.

◆ As each word is introduced in the lesson, write the word on the board and discuss its meaning.

◆ List the words on a Math Word Wall for students to see. As each word is introduced in the lesson, add a picture next to the word on the Word Wall.

◆ Use the vocabulary words regularly when teaching lessons, and encourage students to use the words in their discussions.

Games

Below are suggested Unit 12 game adaptations. For more information about implementing games in a differentiated classroom, see pages 20–25 of this handbook.

Game: *First to 100*

Skill Practiced: Solve open number sentences. [Patterns, Functions, and Algebra Goal 2]

Modification	Purpose of Modification
Players use the sum of their dice instead of the product. Play ends after ten rounds. The player with the higher score wins.	Students solve open number sentences. [Patterns, Functions, and Algebra Goal 2]
Players make a 2-digit number from the roll of the dice. If the final answer is a multiple of 3 or of 4, they get ten bonus points. Play ends after ten rounds. The player with the highest score wins.	Students use place-value concepts to make and solve open number sentences. [Patterns, Functions, and Algebra Goal 2]

Game: *Spoon Scramble*

Skill Practiced: Find equivalent names for rational numbers. [Number and Numeration Goal 4]

Modification	Purpose of Modification
Each player makes a name-collection box for a fraction or a whole number. Using a pencil, each player makes a set of *Spoon Scramble* cards using four expressions from their boxes. Switch sets of cards between groups.	Students find equivalent names for rational numbers. [Number and Numeration Goal 5]
Each player makes a set of cards for a mixed number, including a fraction, a decimal, a percent, and an addition or subtraction number sentence name. Switch sets of cards between groups.	Students find fraction, decimal, and percent names for mixed numbers. [Number and Numeration Goal 5]

Game: *Coordinate Search*

Skill Practiced: Plot points on a coordinate grid. [Measurement and Reference Frames Goal 4]

Modification	Purpose of Modification
Provide a "hint" for players that all islands have a corner pointing down.	Students plot points on a coordinate grid. [Measurement and Reference Frames Goal 4]
Have players write out their plan before they solve the puzzle. When they have found all of the islands, have them make up a similar "search." They can switch grids and solve each others'. Give them the option of using a 4-quadrant grid from –5 to 5 on each axis.	Students plot points on a coordinate grid and write clues for locating points. [Measurement and Reference Frames Goal 4]

Math Boxes

Suggestions for using Math Boxes to meet individual needs begin on page 26 of this handbook. There are blank masters for Math Boxes on pages 136–141.

Using Part 3 of the Lessons

Use your professional judgment, along with assessment results, to determine whether the whole class, small groups, or individual students might benefit from these Unit 12 activities. Consider using the Part 3 Planning Master found on page 154 of this handbook to record your plans.

Readiness Activities

Lesson	Activity	Purpose of Activity
12◆1	**Make factor rainbows.**	Gain experience finding factors of a number. [Number and Numeration Goal 3]
12◆2	**Determine whether or not a theoretical probability can be calculated for each event in a list.**	Gain experience with the language of probability. [Data and Chance Goal 3]
12◆3	**Analyze a pictograph to solve ratio problems.**	Gain experience representing ratios in a variety of ways. [Operations and Computation Goal 7]
12◆4	**Write ratios as fractions in simplest form.**	Explore the relationship between ratios and equivalent fractions. [Number and Numeration Goal 5]
12◆6	**Write ratios using fractions, decimals, and percents.**	Gain experience identifying equivalent names for ratios. [Number and Numeration Goal 5]
12◆7	**Make line plots with stick-on notes and identify data landmarks.**	Gain experience with line plots and statistical landmarks. [Data and Chance Goals 1 and 2]

English Language Learners Support Activities

Lesson	Activity	Purpose of Activity
12◆1	**Make factor-tree posters to illustrate words related to *factor trees.***	Use student-made posters as a visual reference for a new term. [Number and Numeration Goal 3]
12◆2	**Add *equally likely* to the Math Word Bank.**	Make connections among and use visuals to represent terms. [Data and Chance Goal 3]
12◆8	**Draw a Venn diagram listing similarities and differences between *ratios* and *rates.***	Use a graphic organizer to describe characteristics of the terms. [Operations and Computation Goal 7]

Enrichment Activities

Lesson	Activity	Purpose of Activity
12•1	**Use a division method to find prime factorizations.**	Apply understanding of prime factorization. [Number and Numeration Goal 3]
12•2	**Read *Jumanji* and discuss the probabilities for events in the story.**	Explore probability. [Data and Chance Goals 3 and 4]
12•3	**Write a story describing what life would be like if things were increased or decreased by a factor of ten.**	Apply understanding of ratio comparisons and magnitude. [Number and Numeration Goal 6]
12•5	**Use equivalent fractions, quick common denominators, and cross multiplication to solve ratio number stories.**	Explore using cross multiplication to solve ratio problems. [Operations and Computation Goal 7]
12•6	**Research and compare heart rates of animals and humans.**	Explore solving rate problems. [Operations and Computation Goal 7]
12•7	**Collect and compare real exercise data to investigate relationships between heart rate and calories expended.**	Apply understanding of making predictions from data. [Data and Chance Goal 2]
12•8	**Read *If You Hopped Like a Frog,* and discuss the questions at the end of the story.**	Explore ratios. [Operations and Computation Goal 7]

Extra Practice Activities

Lesson	Activity	Purpose of Activity
12•1	**Use factor trees to find common denominators and least common multiples.**	Practice making factor trees to find prime factorizations. [Number and Numeration Goal 3]
12•3	**Read ratios in problem situations and write them in equivalent forms.**	Practice reading and writing ratios. [Number and Numeration Goal 5]
12•4	**Use square tiles to solve a set of ratio problems.**	Practice solving ratio problems. [Operations and Computation Goal 7]
12•5	**Write and solve ratio number stories.**	Practice solving ratio number stories. [Operations and Computation Goal 7]
12•8	**Use data about musical instruments to solve ratio problems.**	Practice using data to solve ratio problems. [Operations and Computation Goal 7]

Looking at Grade-Level Goals

Everyday Mathematics develops concepts and skills over time. Below is a chart showing where the Grade-Level Goals emphasized in this unit are addressed throughout the year. Use the chart to help you determine which Maintaining Concepts and Skills activities on page 133 to utilize to ensure that students continue working toward these Grade-Level Goals.

● Grade-Level Goal is taught.
◍ Grade-Level Goal is practiced and applied.
○ Grade-Level Goal is not a focus.

Grade-Level Goals Emphasized in Unit 12	Unit											
	1	2	3	4	5	6	7	8	9	10	11	12
[Number and Numeration Goal 5] Use numerical expressions to find and represent equivalent names for fractions, decimals, and percents; use and explain multiplication and division rules to find equivalent fractions and fractions in simplest form; convert between fractions and mixed numbers; convert between fractions, decimals, and percents.	◍	◍	○	◍	●	●	◍	●	◍	○	◍	◍
[Number and Numeration Goal 6] Compare and order rational numbers; use area models, benchmark fractions, and analyses of numerators and denominators to compare and order fractions and mixed numbers; describe strategies used to compare fractions and mixed numbers.	◍	◍	◍	◍	●	◍	●	●	◍	○	○	◍
[Operations and Computation Goal 7] Use repeated addition, arrays, area, and scaling to model multiplication and division; use ratios expressed as words, fractions, percents, and with colons; solve problems involving ratios of parts of a set to the whole set.	●	◍	○	○	◍	○	◍	●	◍	●	○	◍
[Data and Chance Goal 4] Predict the outcomes of experiments, test the predictions using manipulatives, and summarize the results; compare predictions based on theoretical probability with experimental results; use summaries and comparisons to predict future events; express the probability of an event as a fraction, decimal, or percent.	◍	●	○	○	◍	●	◍	○	◍	○	○	◍

Maintaining Concepts and Skills

After completing the curriculum, here are several suggestions for maintaining and practicing concepts and skills.

Number and Numeration Goal 5

- Have students play *Fraction Capture* and *Spoon Scramble.*
- Have students write ratios in simplest form. See the Readiness activity in Lesson 12-4 for more information.
- Use the Name-Collection Boxes master on page 147 of this handbook to create practice problems.

Number and Numeration Goal 6

- Have students play *Build-It* and *High-Number Toss.*
- Use the "What's My Rule?" master on page 146 of this handbook to create practice problems in which the rules are written as relationships between numbers, for example, 100 more or 1,000 less.

Operations and Computation Goal 7

- Have students analyze pictograph ratio problems. See the Readiness activity in Lesson 12-3 for more information.

Data and Chance Goal 4

- Have students conduct probability experiments with spinners and coin tosses. Have them design the experiment, predict the outcome, and test their predictions. Then have them report on their experiments. For example, they predict that $\frac{1}{4}$ of the time they will get 2 HEADS when they flip a coin 2 times. They give the reasoning behind their prediction and then test the prediction by repeating the experiment 100 times.

Assessment

See page 144 in the *Assessment Handbook* for modifications to the written portion of the Unit 12 Progress Check.

Additionally, see pages 145–149 for modifications to the open-response task and selected student work samples.

Masters

The masters listed below provide additional resources that you can customize to meet the needs of a diverse group of learners. The templates, pages 136–153, include additional Math Boxes problem sets, more practice with program routines, and support for language development. Use the Part 3 Planning Master, page 154, to record information about your differentiation plan.

Contents

Name Date Time

Math Boxes A

1.

2.

3.

4.

Name Date Time

Math Boxes B

1.	2.
3.	4.
5.	6.

Name Date Time

Math Boxes C

1.

2.

Rule

in	out

3.

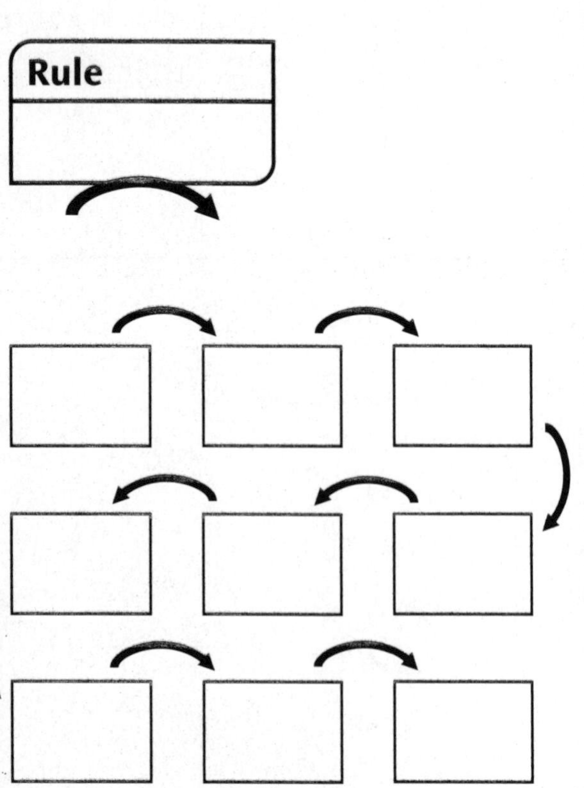

4. Complete the number-grid puzzles.

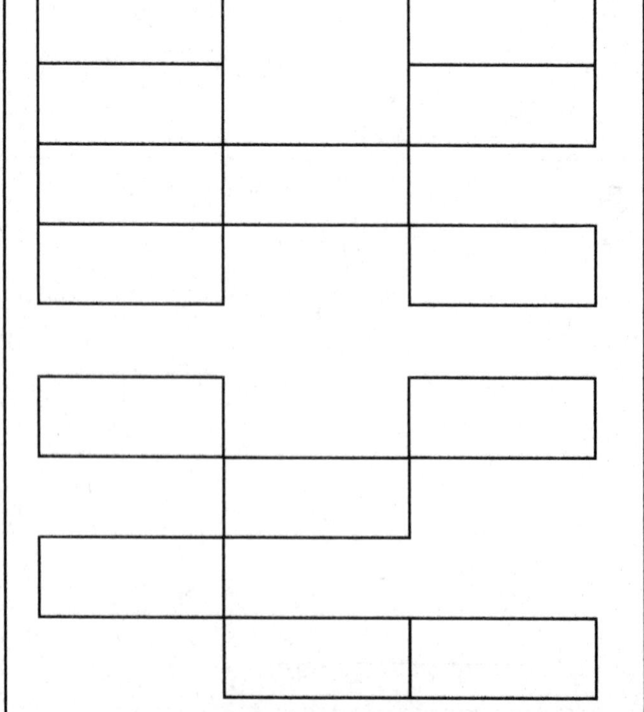

Name Date Time

Math Boxes D

1. Write the numbers in scientific notation.

   a. ______________

   ____

   b. ______________

   ____

2. List all the factors of ________.

3. Write and solve a multiplication problem.

4. Draw a ________° angle.

5. Plot the points on the grid.

   A (_____,_____) *B* (_____,_____)

   C (_____,_____) *D* (_____,_____)

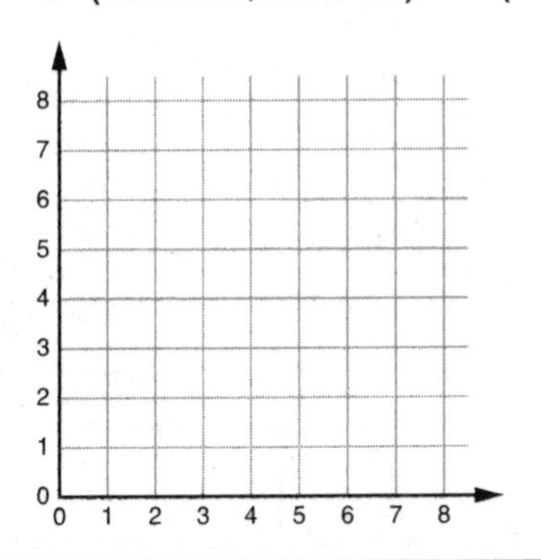

6. Complete the "What's My Rule?" table and state the rule.

Rule

in	out

Name ____________ Date ____________ Time ____________

Math Boxes E

1. Write two equivalent fractions for ____________.

____________ ____________

2. Make a factor tree for ____________.

3.

4. Write the numbers in order from least to greatest.

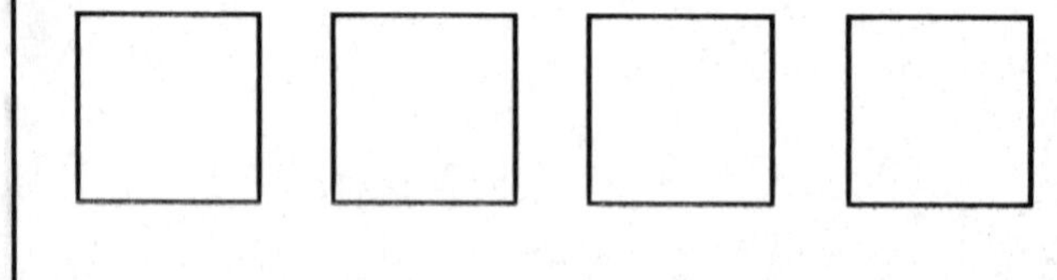

______ ______ ______ ______

5. Write < , > , or =.

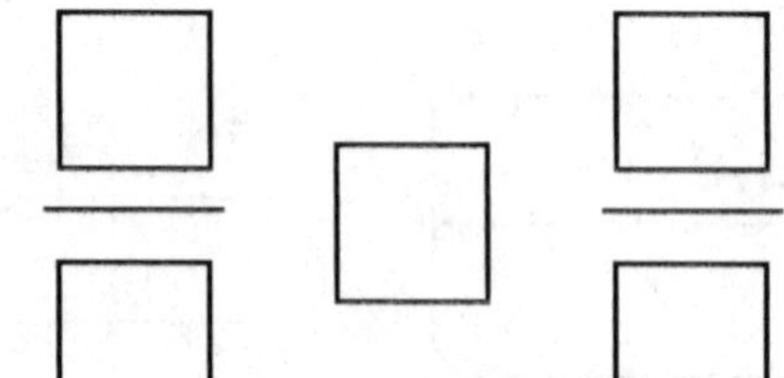

6. Complete.

a. ______ * ______ = ______

b. ______ = ______ * ______

c. ______ = ______ * ______

d. ______ * ______ = ______

Name _______________ Date _______________ Time _______________

Math Boxes F

1. Round ____________________
to the nearest

a. hundred. ____________________

b. thousand. ____________________

c. hundred thousand. ____________________

2. Write and solve a division number story.

3. Write and solve a multiplication number story.

4. A store is giving a __________% discount on all items. Find the sale price for each item.

Item	Regular Price	Sale Price

5. Find the median and mean for each data set.

a. ___, ___, ___, ___, ___

median ____________________

mean ____________________

b. ___, ___, ___, ___, ___, ___

median ____________________

mean ____________________

6. Rename each fraction as a mixed number or a whole number.

a. 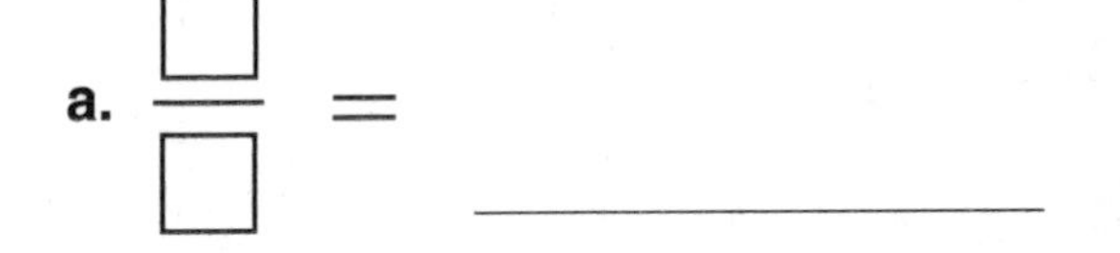

b. 

c. 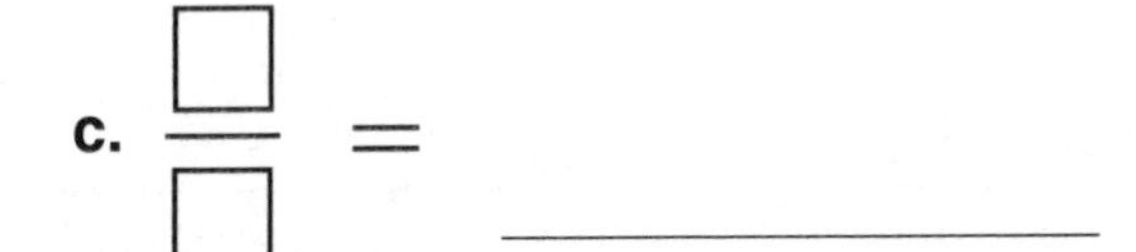

Name Date Time

Math Word Bank A

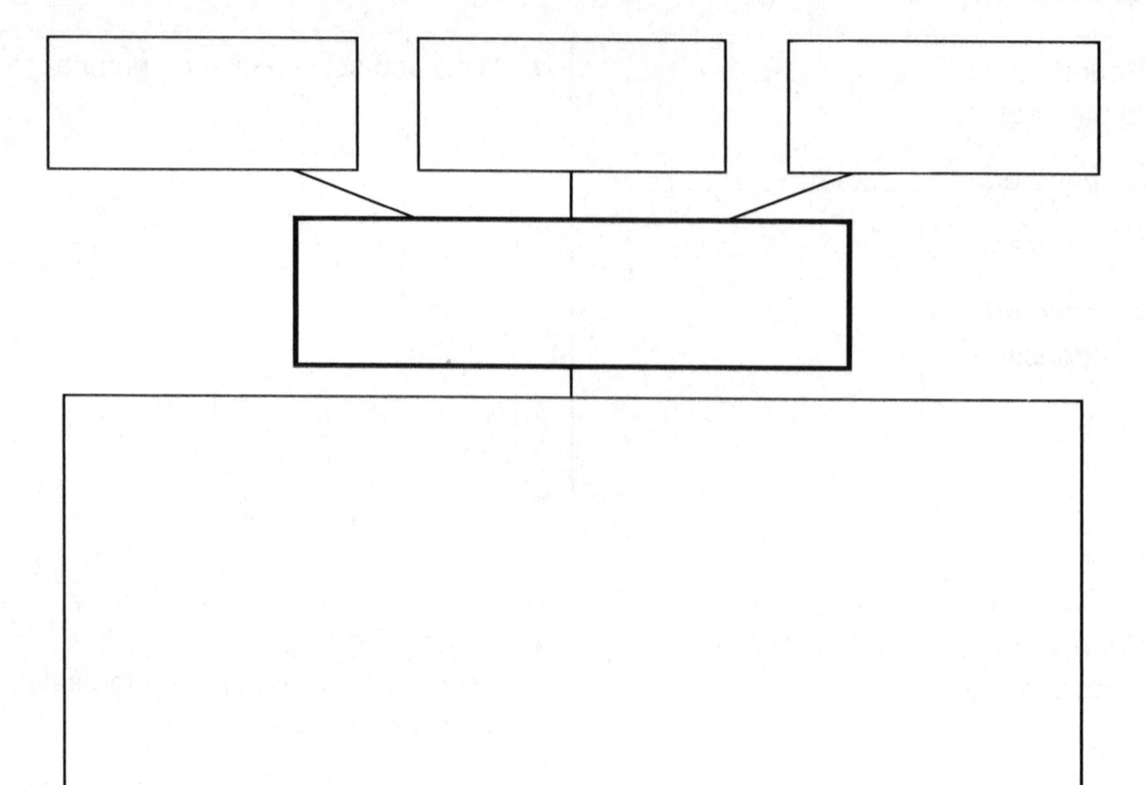

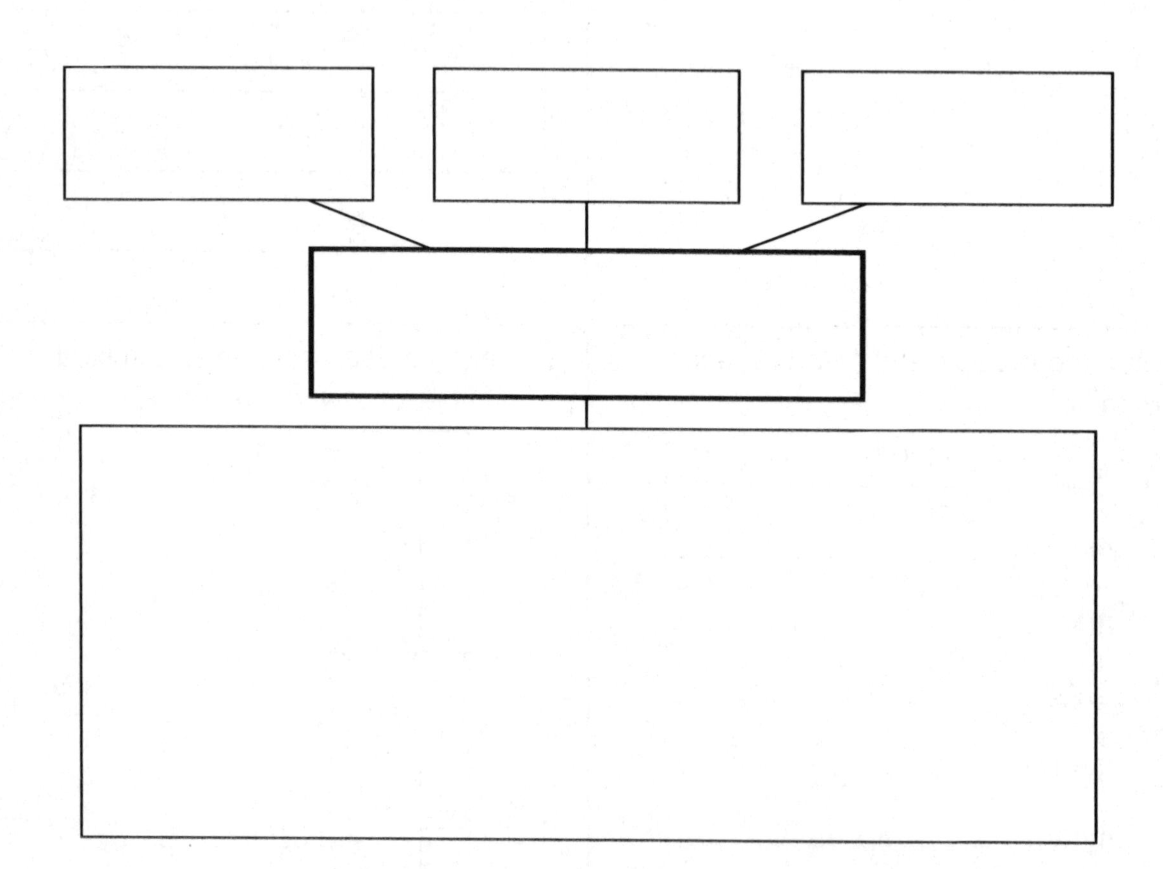

Name Date Time

Math Word Bank B

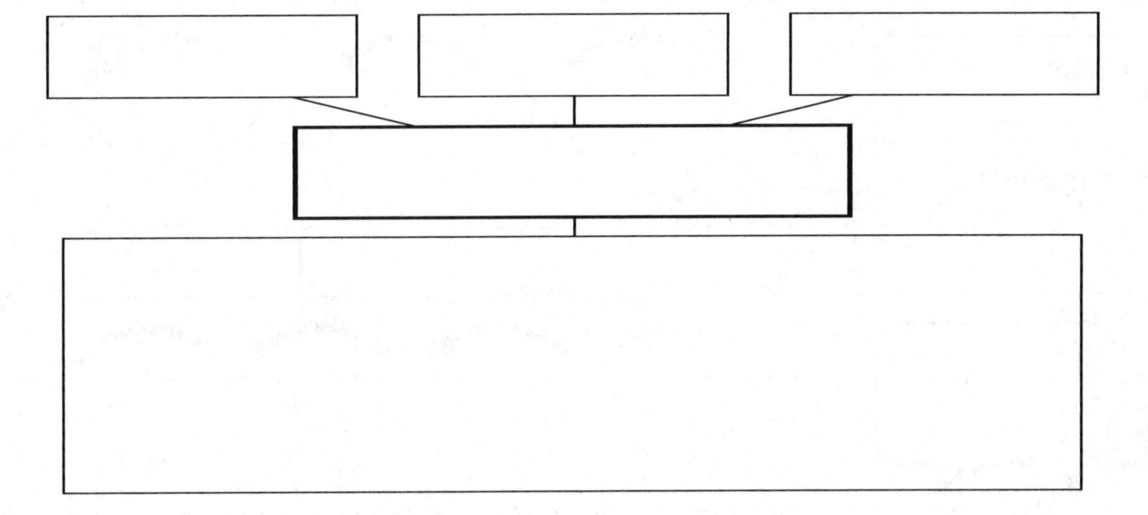

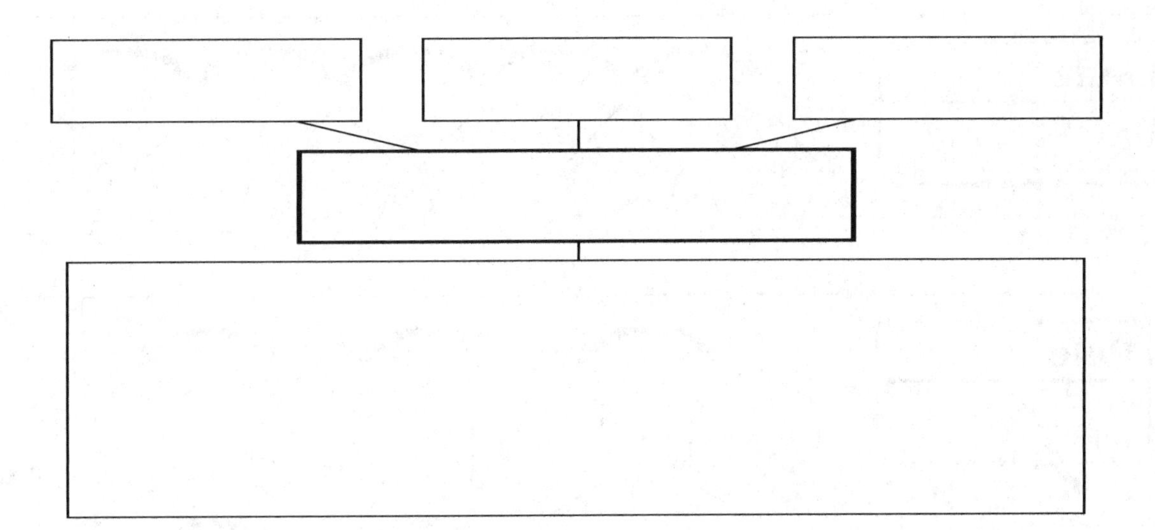

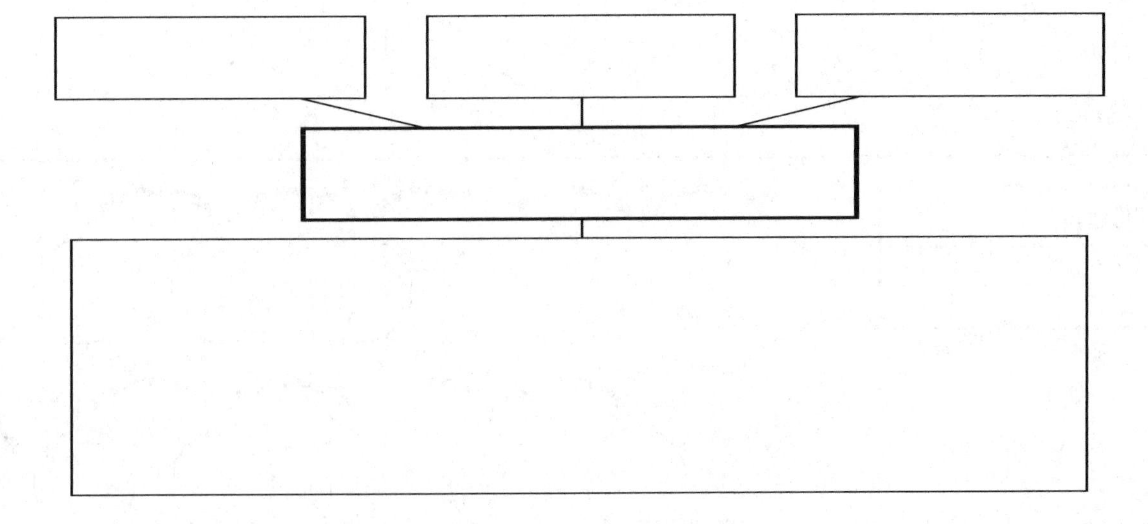

Name Date Time

Frames and Arrows A

1.

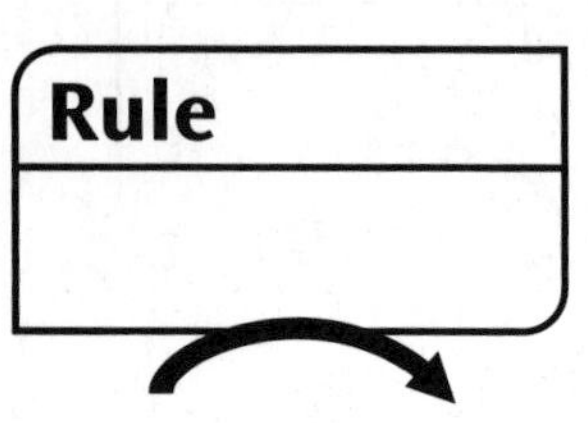

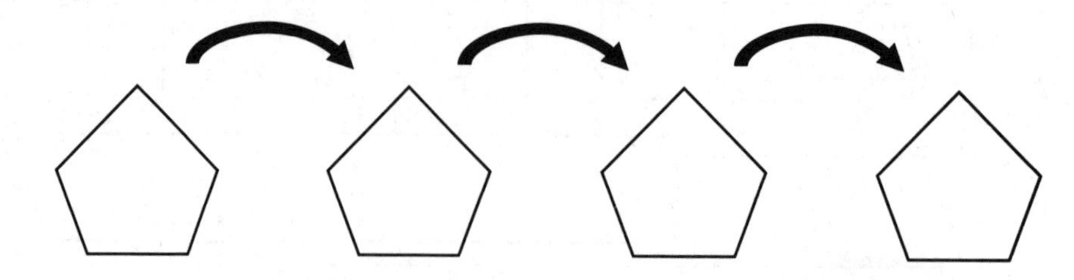

2.

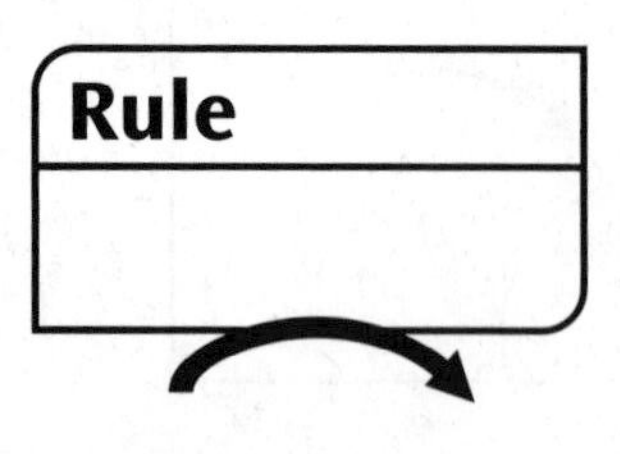

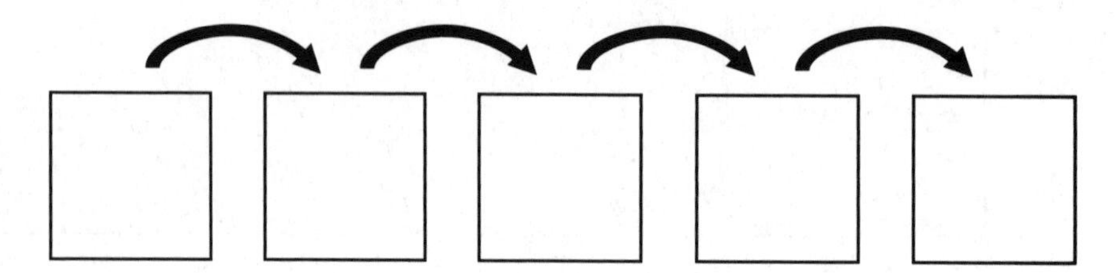

3.

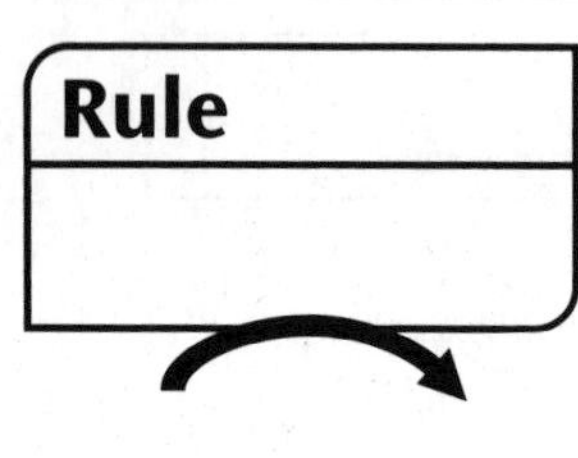

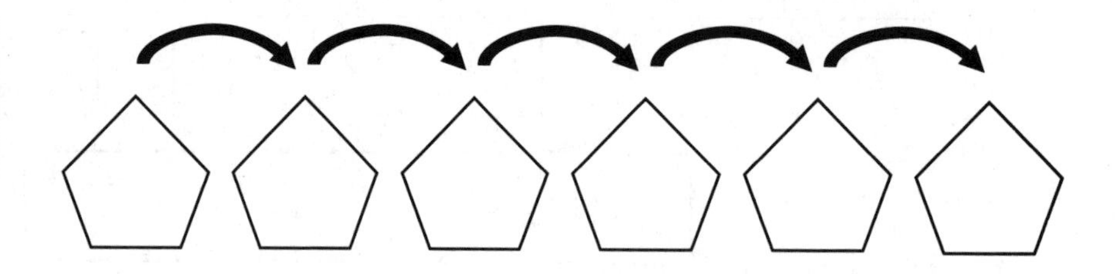

4.

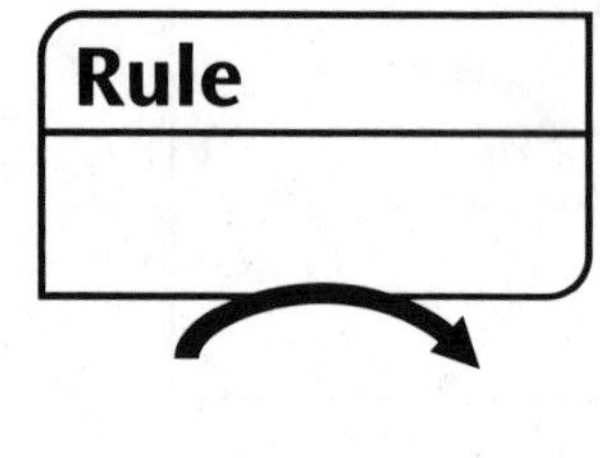

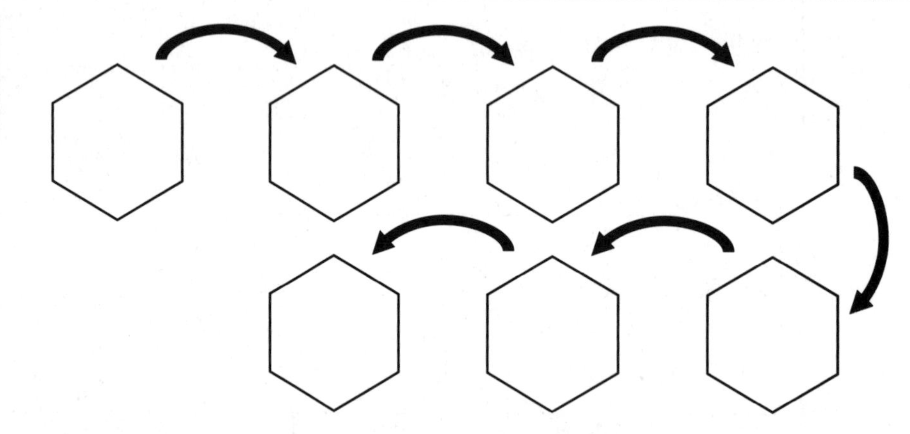

5.

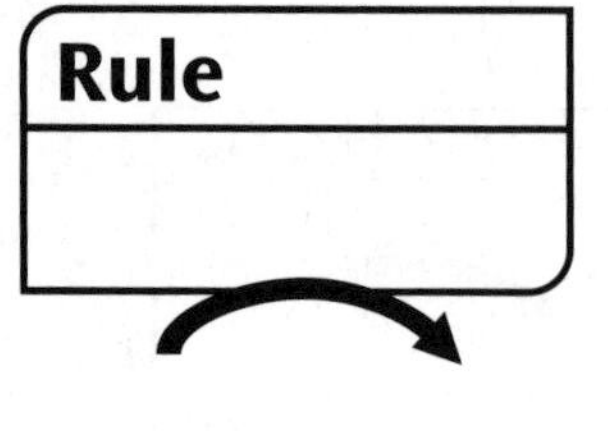

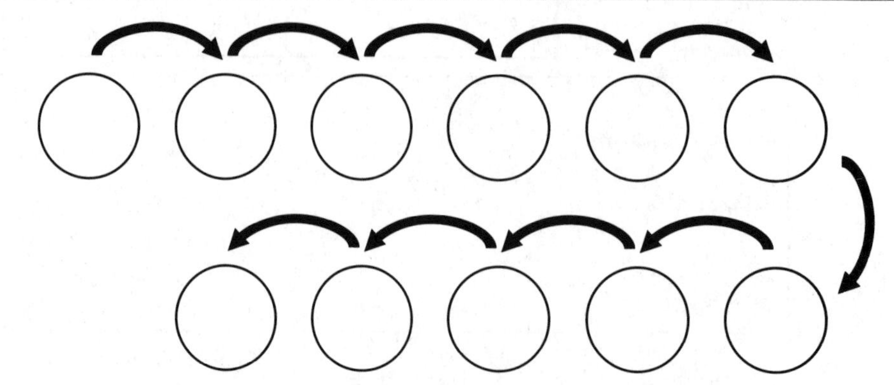

Name Date Time

Frames and Arrows B

1.

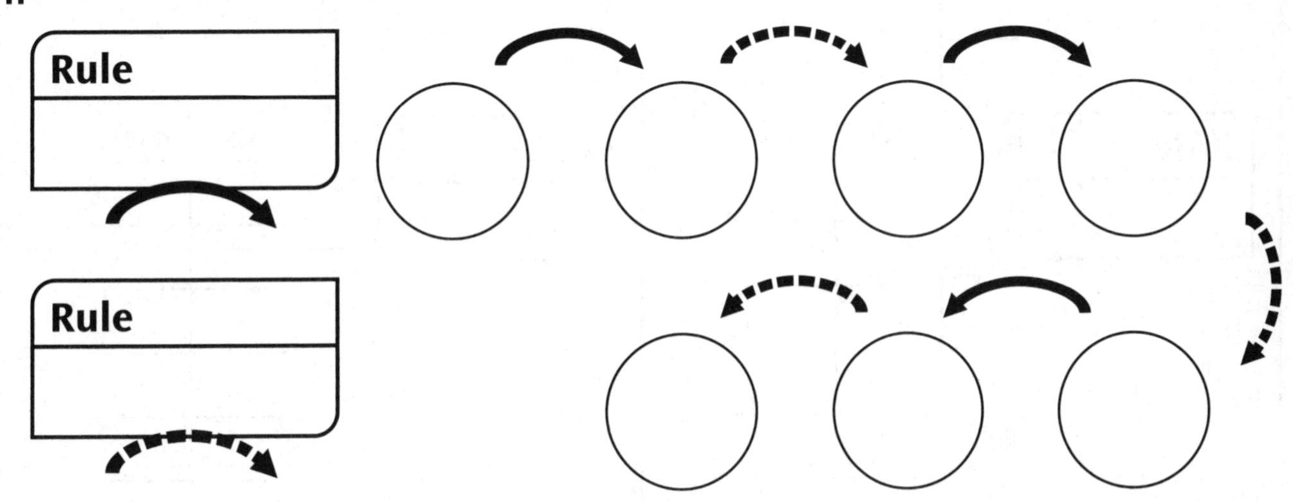

2.

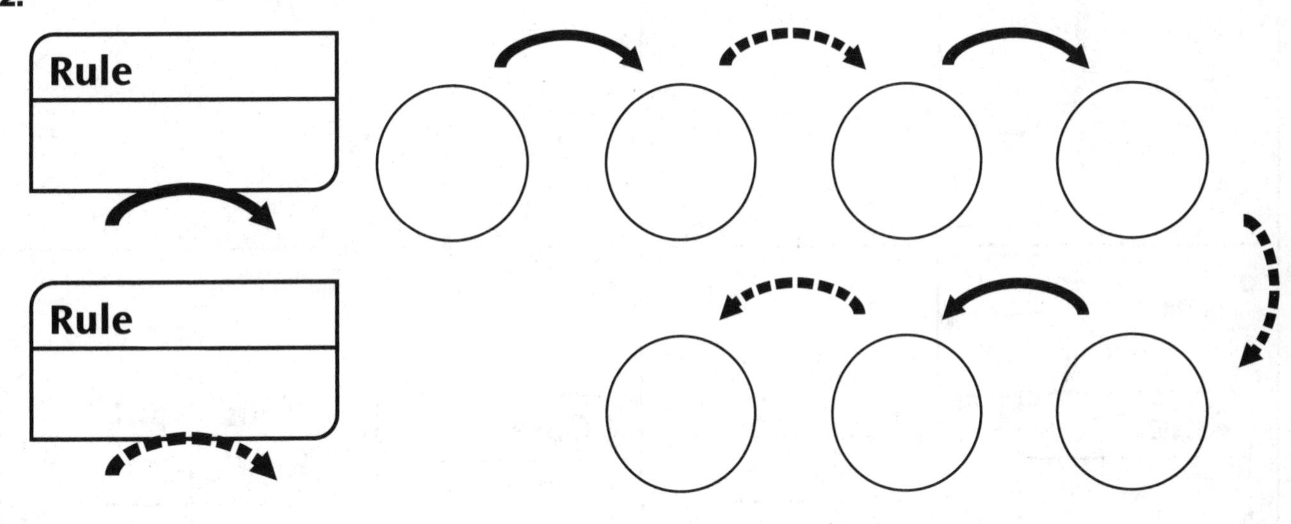

3.

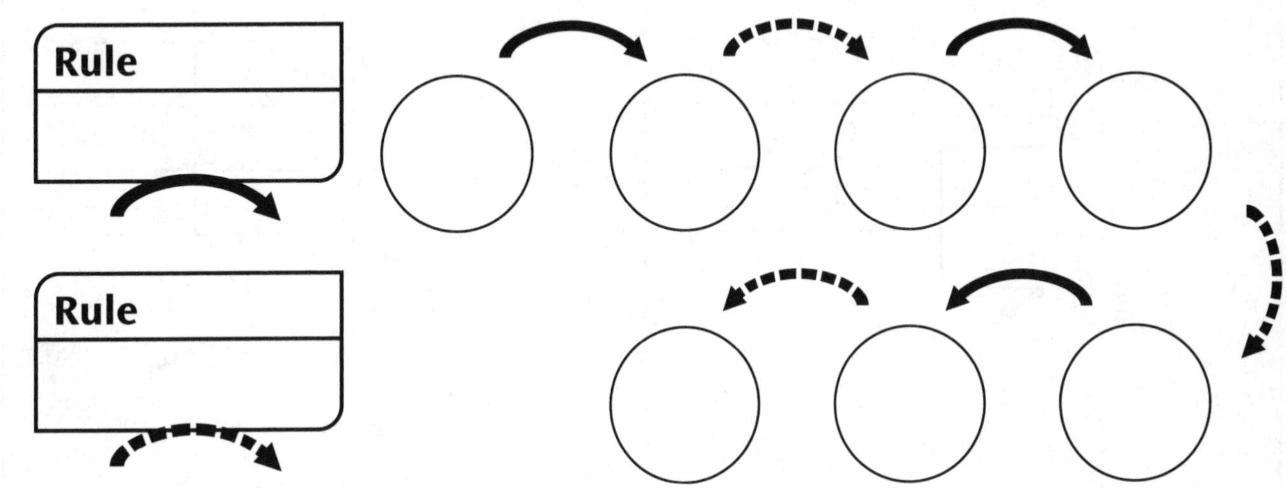

Name Date Time

"What's My Rule?"

1.

Rule

in	out

2.

Rule

in	out

3.

Rule

in	out

4.

Rule

in	out

Name Date Time

Name-Collection Boxes

1.

2.

3.

4.

Name Date Time

Number Grid

−9	−8	−7	−6	−5	−4	−3	−2	−1	0
1	2	3	4	5	6	7	8	9	10
11	12	13	14	15	16	17	18	19	20
21	22	23	24	25	26	27	28	29	30
31	32	33	34	35	36	37	38	39	40
41	42	43	44	45	46	47	48	49	50
51	52	53	54	55	56	57	58	59	60
61	62	63	64	65	66	67	68	69	70
71	72	73	74	75	76	77	78	79	80
81	82	83	84	85	86	87	88	89	90
91	92	93	94	95	96	97	98	99	100
101	102	103	104	105	106	107	108	109	110

−9	−8	−7	−6	−5	−4	−3	−2	−1	0
1	2	3	4	5	6	7	8	9	10
11	12	13	14	15	16	17	18	19	20
21	22	23	24	25	26	27	28	29	30
31	32	33	34	35	36	37	38	39	40
41	42	43	44	45	46	47	48	49	50
51	52	53	54	55	56	57	58	59	60
61	62	63	64	65	66	67	68	69	70
71	72	73	74	75	76	77	78	79	80
81	82	83	84	85	86	87	88	89	90
91	92	93	94	95	96	97	98	99	100
101	102	103	104	105	106	107	108	109	110

Copyright © Wright Group/McGraw-Hill

Name Date Time

Venn Diagram A

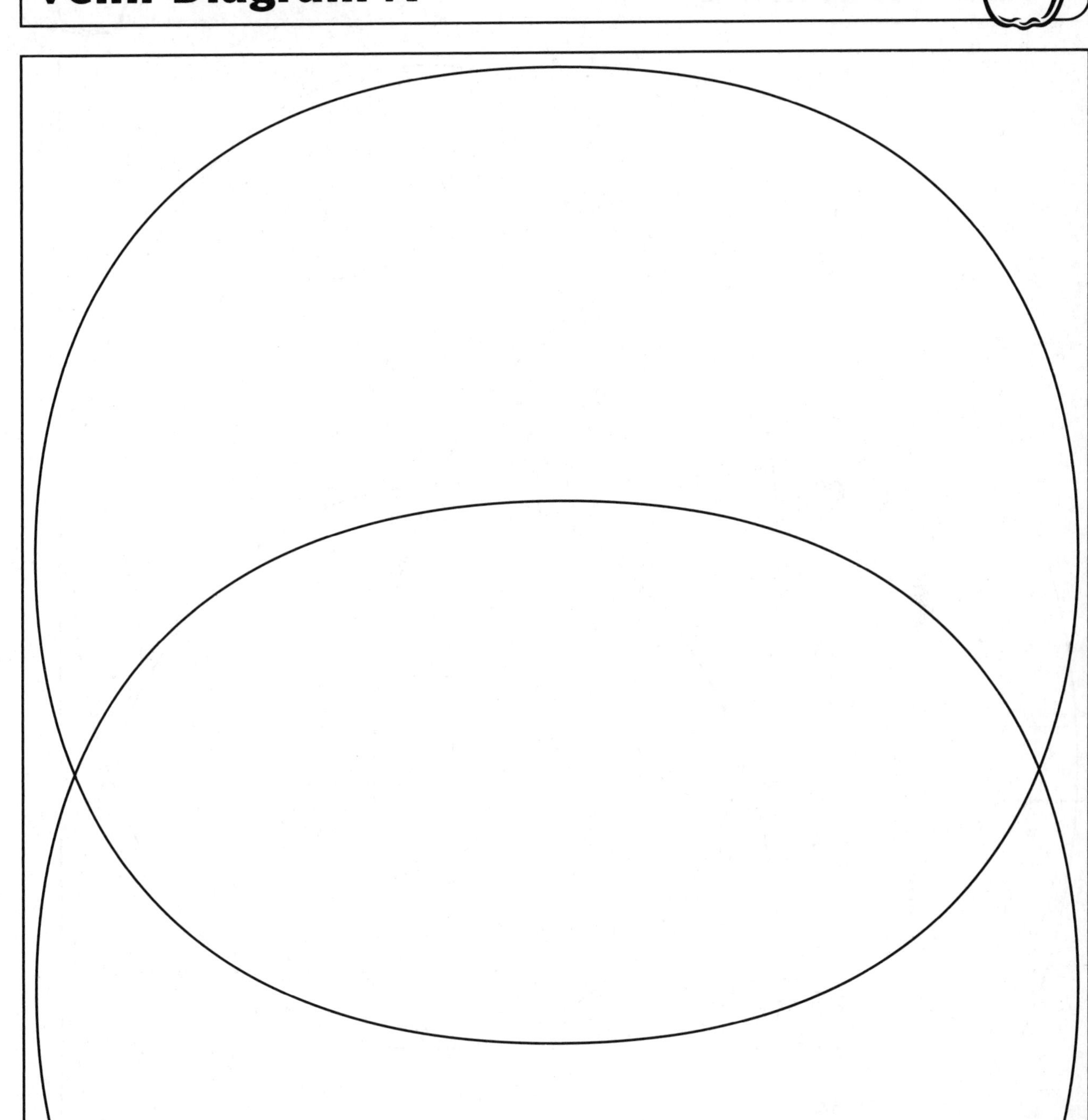

Name Date Time

Venn Diagram B

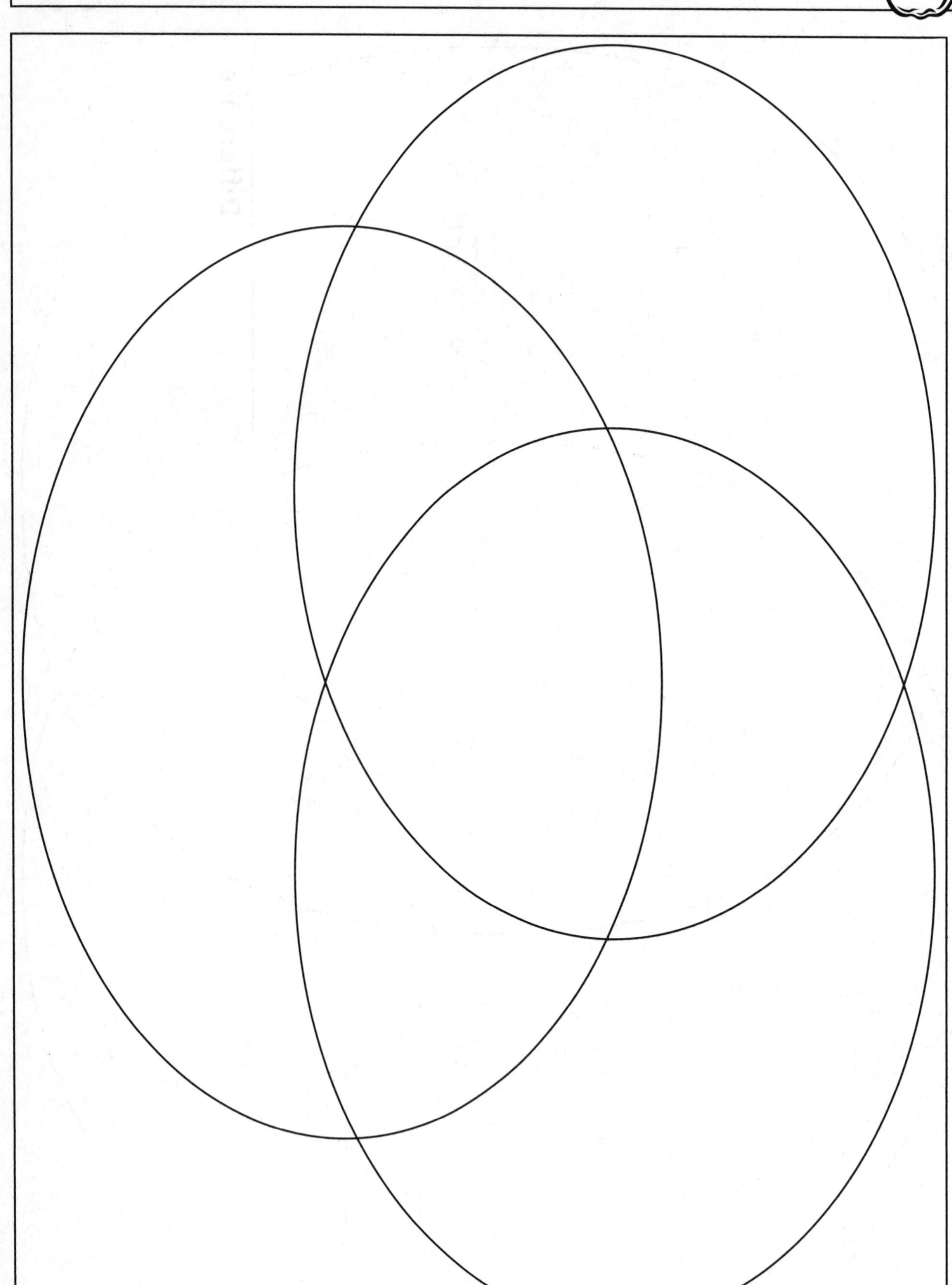

Name Date Time

Situation Diagrams for Number Stories

Change	Parts-and-Total	Comparison
Start / Change / End	Total / Part / Part	Quantity / Quantity / Difference
Start / Change / End	Total / Part / Part	Quantity / Quantity / Difference

Name ______________________ Date ______________________ Time ______________________

Multiplication/Division Diagrams

_______	_______ per _______	_______ in all

_______	_______ per _______	_______ in all

_______	_______ per _______	_______ in all

Name Date Time

Study Link

Part 3 Planning Master

Lesson	Readiness	Enrichment	Extra Practice	ELL Support

Resources

Recommended Reading

Baxter, Juliet A., John Woodward, and Deborah Olson. 2001. "Effects of Reform-Based Mathematics Instruction on Low Achievers in Five Third-Grade Classrooms." *The Elementary School Journal* 101 (5): 529–547.

Garnett, Kate. 1998. "Math Learning Disabilities." LD OnLine. www.ldonline.org (accessed Jan. 19, 2004).

Johnson, Dana T. 2000. "Teaching Mathematics to Gifted Students in a Mixed-Ability Classroom." Reston, Va.: ERIC Clearinghouse on Disabilities and Gifted Education.

Lock, Robin H. 1997. "Adapting Mathematics Instruction in the General Education Classroom for Students with Mathematics Disabilities." LD OnLine. www.ldonline.org (accessed Dec. 15, 2009).

Tomlinson, Carol Ann. 1999. *The Differentiated Classroom: Responding to the Needs of All Learners.* Alexandria, Va.: Association for Supervision & Curriculum Development.

Usiskin, Zalman. 1994. "Individual Differences in the Teaching and Learning of Mathematics." Chicago, Ill.: UCSMP Newsletter 14 (Winter).

Villa, Richard A., and Jacqueline S. Thousand, eds. 2005. *Creating an Inclusive School.* Alexandria, Va.: Association for Supervision & Curriculum Development.

http://everydaymath.uchicago.edu/

References

Gregory, Gayle H. 2003. *Differentiated Instructional Strategies in Practice: Training, Implementation, and Supervision.* Thousand Oaks, Calif.: Corwin Press.

Robertson, Connie, ed. 1998. *Dictionary of Quotations (Wordsworth Reference Series).* 3rd Rev. Edition. Hertfordshire, UK: Wordsworth Editions Ltd.

Tomlinson, Carol Ann. 2003. "Deciding to Teach Them All." *Educational Leadership* 61 (2): 6–11.